The Popular Science

Router Book

The Popular Science

Router Book

by Doug Geller

Drawings by **Michael Goodman**
Photographs by **Michael Price**

 SEDGEWOOD® PRESS, New York

For my parents

Published by
   Popular Science Books
   Sedgewood® Press
   750 Third Avenue
   New York, New York 10017

Distributed by Meredith Corporation, Des Moines, Iowa.

Produced by Soderstrom Publishing Group Inc.
Designed by Jeff Fitschen

ISBN: 0-696-11092-X

Library of Congress Cataloging-in-Publication Data

Geller, Doug.
   The Popular science router book.

   Includes index.
   1. Routers (Tools) 2. Woodwork. I. Popular
science (New York, N.Y.) II. Title. III. Title:
Router book.
TT203.5.G44   1987   684′.083   87-14842
   20 19 18

Contents

Preface

The router is a wondrous tool. Friends who know nothing about it sometimes ask me what the router can do, and I begin the usual litany—"It cuts dadoes, dovetails, mortises, tenons, makes moldings, cuts perfect circles, straightens the edges of curved boards, cuts miters and spline grooves, carves signs, trims laminates . . ." and I have to stop myself because I realize I sound like one of those late-night television commercials selling some magic vegetable chopper. But the fact is that the router really does do all of these things and does them very well—so well and with such accuracy that one is tempted to forget how simple a machine it really is.

As a professional cabinetmaker, I use the router every day. The uses to which I put it are often repetitive and mundane—cutting rabbets for cabinet backs, cutting innumerable grooves, making drawers—and I begin to take it for granted. Then an architect's drawing finds its way into my shop with a design, or just a detail, that defies comprehension, let alone solution. We all stare blankly at the drawing. Then someone picks up the router, and before you know it, a jig is being made. A bit is inserted, and we make a test cut. The bit is changed and the jig modified. The router solves problems like no other machine in the shop.

However, the use of the router must be kept in proper perspective. I think that there is a tendency to want the one machine that can do it all. While it is true that the router can, if set up properly, simulate most of the operations that cabinetmakers perform on machines such as the table saw, radial arm saw, and jointer, the router cannot be viewed as something to replace these machines. Yes, for example, the router can produce a miter cut, but most miters are better made on the table saw. Yes, the router can be set up to function like a jointer, but the jointer was designed to perform its function with ease and simplicity. At the same time, on some occasions mitering and jointing are best done with the router. And, of course, the router can do many things that other machines cannot do. Its versatility in this regard is unparalleled. While some professional woodworking shops might get by without a drill press or a radial arm saw, I doubt that there is a professional shop anywhere that operates without a router. Even an Amish cabinetmaker I know cannot conceal his enthusiasm for his air-driven router.

For all its magical qualities, however, the router apparently remains a mystery to many or at the very least is not used to its fullest advantage. The router's piercing sound alone is enough to discourage some people from attempting to learn its full capabilities. I can't help but think that many a dust-covered router has been set aside by its intimidated owner. Yet I have attempted both to show the beginner how simple a tool the router actually is and to demonstrate for the advanced woodworker many more sophisticated routing techniques.

My approach here is not encyclopedic. I have omitted techniques such as "turning" and "carving" because the router is not wholly appropriate for these purposes. Although the router can, through the use of commercially available accessories, "turn" and "carve," to use it thus somehow offends my more traditional sensibilities. Along this line, I have omitted discussion of the many gadgets aimed at the home hobbyist. If I wrote about them, it would be with little enthusiasm. Instead, I share the techniques that I have inherited and those that I have developed during my career as professional woodworker. If you can master all of the procedures set forth in this book, there will be few routing problems you cannot solve. Creative woodworking requires the ability to take a familiar technique and use it in a new way.

When I think about the router and its vast ability, I remember a young fellow who studied cabinetmaking with me a few years ago. Paul was intelligent, had good hands, and was keenly interested in learning, but had the disadvantage of being blind. He did not think blindness was a disadvantage, though. With his braille ruler and quiet determination, he produced his first cabinet: a beautiful cherry bookcase with two doors and a drawer.

I had not seen him for almost a year when he showed up one day, proudly carrying a book under his arm. It was his woodworking portfolio—page after page of beautiful color photographs of the furniture he had built. There were stereo cabinets, bookcases, tables, and dressers, in both solid and veneered woods. I was impressed! I asked him where he had set up shop and what tools he had bought. With a half-concealed smile, he answered, "In a friend's garage. But I only had one real tool—a router!"

Doug Geller

Acknowledgments

Many people have helped make this book possible. First and foremost, I would like to thank Michael Price for his great patience and dedication in taking the many photographs and Michael Goodman for his excellent illustrations. Miraculously, two valued friendships have survived this project.

I would also like to thank Bob Cezark at the Fred M. Velepec Company for sharing with me a lifetime of router-bit knowledge and taking me on a tour through the Velepec manufacturing plant. Thanks, too, to Rick Schmidt of Porter-Cable for his unhesitating offering of materials and information, and to the others who helped in this way: Jim Preusser of Dinosaw, Inc.; Glen Docherty of Constantine's; Stuart Ain of HIT Distributors; Vince Pax of Shopsmith, Inc; and Ken Grisley of Leigh Industries Ltd.

Nor can an author's book come about without the help of publishing professionals. First, I would like to thank agent Phyllis Wender, of Rosenstone/Wender, for placing this book with Popular Science Book Club, Grolier Books. Thanks also to publisher John Sill for his encouragement and to producer Neil Soderstrom for his improvement of text and photo prints, and preparation of all other materials for the printer.

A very special thanks to Edith Geller for typing the manuscript and for her critical eye in reading it. And I am grateful to my wonderful wife Sandi, for her great patience with me, even though she doesn't consider writing real work. Thanks also to friends John Gregory and Patrick Thaddeus for their enthusiastic technical help.

Finally, I express my deepest appreciation to Gene Liberty, who knows the value of friendship and the joy that is woodworking.

1 Router Basics

For the best understanding of a router and full exploitation of all its possibilities, begin by thoroughly examining one. Here I recommend that you start your examination away from your workshop so you will not be tempted to do any cutting immediately. If you relax and take the time to get to know your router, you will not find yourself later mumbling "so that's what that little nut was for!"

As I stated in the Preface, a router cuts dadoes, dovetails, mortises and tenons. It also makes moldings, cuts perfect circles, straightens the edges of curved boards, cuts miters and spline grooves, carves signs, trims laminates, and more. But before you can take advantage of all the router's possibilities, you must know your tool well.

THE SIMPLE FACTS

The router consists of a motor, an adjustable base, and handles for gripping the tool (**fig. 1-1**). The motor, which for general cabinetwork has a horsepower of between one and three, turns a spindle or shaft. At the end of the spindle is a collet, which is essentially the same as a chuck on an electric drill. You insert your bits or cutters (terms we will use interchangeably) into this collet. There are literally dozens of bits from which to choose; which partly explains why the router can make so many different types of cuts. By adjusting the router base (that is, moving it up or down), you can regulate the depth of cut. Tighten the thumbscrew (or thumb clamp), and the base is locked in position (**fig. 1-2**).

1-1: The router consists of a motor, an adjustable base, and handles for gripping the tool.

The base also ensures that the router and bit are perpendicular to the work surface.

The router derives its clean and efficient cutting ability not from its horsepower, but from its amazingly high rate of speed—25,000 revolutions per minute on the average model. Compare this with an electric drill, which has a typical speed of 2,500 r.p.m., or one-tenth the speed of a router. The high rate of a router's speed accounts for the smooth, polished surface produced by the bits, while the relatively small horsepower partly explains why you are limited in how deep a cut you can make on any one pass. By comparison, a shaper, which in a certain sense can be construed as a powerful, stationary router, can make very large cuts in one pass by virtue of its more powerful motor and greater horsepower. While the router's rate of speed is what makes it cut smoothly, its horsepower is nonetheless a very important factor in determining its performance. The 2- to 3-h.p. high-performance routers can cut more deeply and faster than their lighter-duty counterparts.

The lightest-duty routers are not called routers at all, but rather, laminate trimmers (1 h.p. or less on the average). As the name implies, they are used for trimming plastic laminates and veneer. On the other end of the continuum are the powerful plunge routers (and some industrial nonplunging types) with between 2 and 3½ horsepower and weights of 10 pounds (4.5 kg) and more. Although these larger routers are becoming increasingly popular, the vast majority of routers sold are in the 1 to 2 horsepower range. A discussion of which router is appropriate for you can be found in the upcoming section "Choosing a Router."

Routers vary not only in terms of horsepower, but also in the shank size of the bits they accept. Routers with horsepower of 1¼ or less will usually accept bits with shanks of ¼ inch (6 mm) and less. Routers of 1¼ h.p. or more will accept bits with ½-inch (12-mm) shanks. Because the majority of bits used have ¼-inch (6-mm) shanks, it is necessary, with the larger routers, to insert a collet adapter into the collet. Some of the very small bits, such as the fine-veining bits, have shanks that are less than ¼ inch (6 mm), for which the appropriate adapters will have to be used. Collets and collet adapters come in a variety of styles. Among the most common are the split-sleeve, the screw-in, and the straight-sleeve (**fig. 1-3**).

The router base, to which the handles or

1-2: The base of this Makita router is locked into position by depressing a thumb clamp. The base of the Porter-Cable router, on the previous page, is tightened by means of a thumbscrew.

1-3: Typical collet adapters. The two at the left are the screw-in type. Two at the right: split-sleeve.

knobs are attached, is usually round, although Makita, for one, makes a square-based router. The diameter will vary from model to model, but the average size is 6 inches. The major advantage of a round base is that, when pushing the router against a guide, the router can be pivoted (intentionally or otherwise) in any direction without affecting the distance between the guide and the cut. However, this presupposes that the bit and base are perfectly concentric. The fact is that most routers are nonconcentric to some fractional degree. For a router with a round base to work properly, not only does the bit have to be concentric to the motor, but the base has to be concentric to the bit. Finally, the sub base has to be in perfect alignment. To test a base for concentricity, run the router against a straightedge, making a shallow, straight cut (**fig. 1-4**). Pivot the router as the cut is being made so that all of the base has, at one point, touched the fence. Measure the distance from the straightedge to the cut: it should be constant all along the length of the cut. If the base and bit are not

concentric, all you can really do is try another base. Failing that, send the router back to the manufacturer.

The base is usually fitted with a nonmarring (phenolic or plastic) sub base. The sub base is attached with several countersunk machine screws and can easily be removed. Remember that the sub base can be perfectly round, but if not positioned properly on the base itself (due to faulty machining at the factory, so that the bit is not in the exact center of the circle), your router will cut as if the base were not a perfect circle. Although rarely a problem, incorrect positioning is something to be aware of. It is a good idea to mark the correct alignment lines on your base and sub base so that the sub base is always placed back in the same position should it be removed. Switching your sub base position can cause nonconcentricity.

Router on-off switches are mounted directly on the motor or, in some cases, right on the base itself. A few manufacturers offer a D-handle model that houses the trigger mechanism (**fig. 1-5**).

1-4: Testing router base for concentricity. With the router pressed tightly against a perfectly straight fence, rout a groove for a length of about 6 inches (15.24 cm). Then turn the handles 90 degrees (clockwise) and rout a few more inches. Repeat until the router is in its original position. Now measure the distance from the fence to the groove at various points along the cut. There should be no deviation in this measurement.

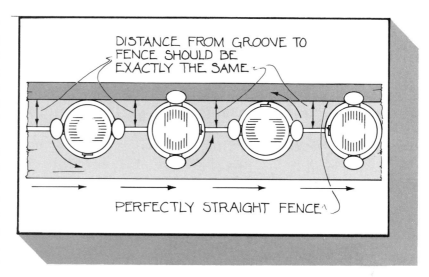

1-5: This D-handle 1½-h.p. router, made by Porter-Cable, is an excellent machine. The vacuum attachment, which attaches to a specially designed base, is a terrific option.

Most routers come equipped with a depth gauge that registers the depth of cut (**fig. 1-7**). To use the gauge, adjust the base so that it is flush with the bottom of the cutter. Set the gauge to zero. As you adjust the depth of cut, it will be registered on the gauge. The depth gauge is not a necessary router feature, although some woodworkers find it a convenience. I prefer to measure the depth of cut with a steel ruler, checking my accuracy with a test cut (**fig. 1-8**). One router model, the Makita 3600B plunge router, has, as a function of a pivoting stopper, the ability to

"remember" two depth settings (**fig. 1-9**). This feature is very useful when making repetitive cuts that require two passes.

Neophyte router owners quickly discover that certain larger-sized bits do not fit through the center hole of the sub base. Rather than enlarging the hole, it is simpler to make an auxiliary sub base. I usually make my sub bases from ¼-inch (6-mm) tempered masonite (smooth on one side). A 1½-inch (36-mm) center hole is sufficient (**fig. 1-6**). Other auxiliary sub bases are discussed in Chapter 11, *Freehand Routing.*

1-6: Substituting an auxiliary sub base. This sub base, made from ¼-inch (6-mm) masonite, has a 1½-inch (38-mm) diameter center hole.

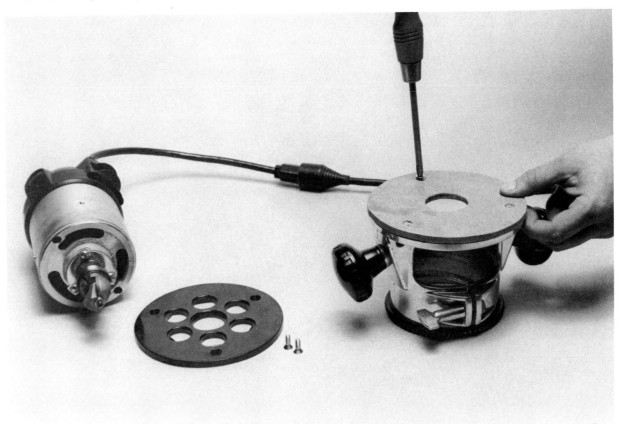

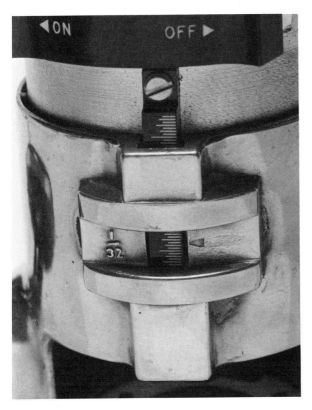

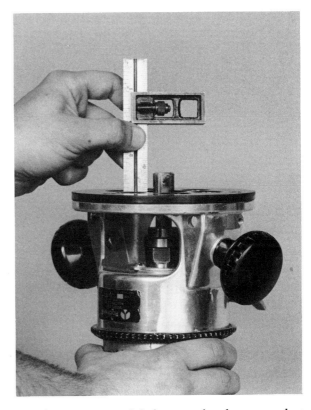

1-7 (left): A typical depth gauge. Most routers, like this Sears model, have a depth gauge that registers depth of cut to within ⅟₃₂-inch (.79 mm). **1-8 (right):** Most woodworkers prefer to measure the depth of cut with a steel ruler and check their accuracy with a test cut.

1-9: This Makita router has a pivoting stop that allows the router to be plunged to one of two preset dimensions. This feature is particularly useful when making repetitive cuts that require two passes.

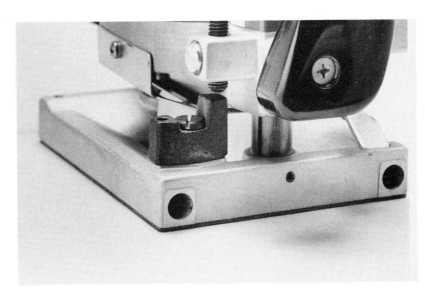

CHANGING A BIT

Although it is not necessary to remove the router base when changing a bit, you will find that, on most models, removal of the router base makes the job easier. On some models the base simply slides off; on other models, it is screwed off.

Router bits are held in the collet (or collet adapter) by a nut that is tightened around it. To remove a bit, this collet nut must be loosened (turned counterclockwise) with a wrench. The spindle, which remains in place, is held steady by a second wrench, although several manufacturers make a spring-loaded locking lever, which, when engaged, prevents the spindle from turning. Only one wrench is needed, therefore, on those models.

Whenever you are changing a router bit, begin by unplugging the machine. There are no exceptions to this rule. Then, lay the router on its side. If your router does not have a locking lever for the spindle, engage the wrench, or use another device to prevent

1-10: Correct removal of a router bit. The wrench resting against the workbench prevents the shaft of the router from turning. The right hand taps the upper nut loose. This controlled tap prevents the fingers from getting pinched.

1-11: Incorrect removal of a router bit. Loosening the upper nut by squeezing the wrenches together will only result in smashed knuckles.

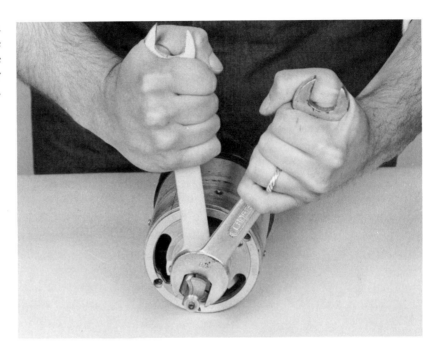

the spindle from turning. Place the second wrench on the upper nut (the one on the collet). With the palm of your hand (not your fingers), tap the wrench down toward the table (**fig. 1-10**). The critical point is to use your palm, so that your fingers will never get pinched between the two wrenches, as so often happens when the wrenches are gripped as shown in **figure 1-11**.

With the upper nut loose, the bit should be free, although many times it will need a little persuasion to "extricate itself," as a Jamaican cabinetmaker once told me. Actually, a stuck bit is a common problem, but a gentle tap on the back of the cutter with a wooden block (never hit the cutting edge of the bit) will usually do the trick. For stubborn cutters, I sometimes use a nut and bolt loosener such as "Liquid Wrench."

To put in a new bit, begin by cleaning the collet adapter and/or the collet. This step is often overlooked. Dust and dirt cause wear, and a worn collet (or adapter) will result in a slipping bit, which could be dangerous. Routers really require very little maintenance, and keeping the collet clean is not

much of a chore. I use the air gun on my compressor to blow out the dust, but a lightweight rag can just as easily be used (**fig. 1-12**). The continual insertion and removal of bits will also cause the collet entrance to wear faster than the rest of the collet. This

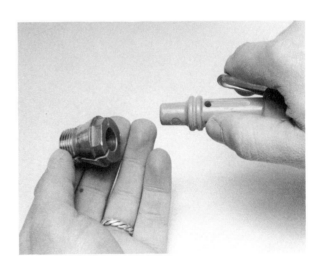

1-12: Keep your collets and collet adapters clean. An air gun is the most effective cleaning tool, but a lightweight rag can also be used.

wear is called *bell mouth*. If the shanks of your bits show excessive or unusual wear marks, you can suspect a worn collet (or adapter). Change it immediately.

Now insert the bit into the collet or adapter. Most manufacturers recommend that the bit be inserted completely and then backed out ⅛ inch (3 mm). This step is taken because most router bits have what is known as a *fillet* just above the cutting edges (**fig. 1-13**). The rounded fillet is wider than the rest of the shank; if it is riding in the collet, then the bit is not being held evenly. Backing the bit out ⅛ inch will ensure that the fillet is in the correct position and that, more importantly, any vibrations caused by the cutting action will not be transferred directly to the motor armature. Tighten the upper nut by squeezing the wrenches (**fig. 1-14**). Notice that in the photograph the router is locked in a vise in an inverted position. Many routers have small, flat flanges (access to the router's brushes), which are perfect surfaces for vise mounting (**fig. 1-15**). If your router does not have such flanges, never lock it in a vise. Such lateral pressure could distort the casing and ultimately ruin the router. An alternative method, then, for tightening the upper nut is shown in **figure 1-16**. Be sure to give the wrenches a very firm squeeze (you need not strain); after all, a flying router bit is one surprise you do not need. Replace the base and position it so that the bit protrudes to the proper depth.

1-13: The rounded surface between the cutting edges and shank of a bit is called a *fillet*. The fillet should not be pushed all the way into the collet, which explains why some manufacturers suggest pushing a bit all the way into a collet and then backing it out about ⅛-inch (3 mm).

1-14 (left): Here, a collet adapter is tightened by squeezing two wrenches together. **1-15 (right):** Many routers have flanges which, for lack of a better term, I call "mounting ears." Your router must have such flanges if you plan to lock it in a vise. Without these flanges, you run the risk of distorting or breaking the router's housing.

1-16: An alternative method for tightening a bit.

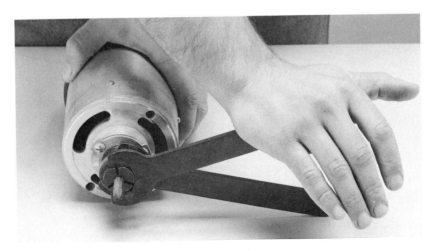

CHOOSING A ROUTER

If you do not already own a router, or do but are not happy with it, this section is for you. The first question to resolve is how many routers you are going to own. This may seem extravagant, but if a bass fisherman can own seven or eight rods, why can't a woodworker own more than one router? Be realistic about your needs. The weekend woodworker can certainly make do with one router. If you are in this category, I would suggest a medium-weight router with a horsepower of no less than 1½. Certainly the manufacturers of 1- and 1¼-h.p. routers would take issue with this. But I have seen too many new wood-

workers struggling with underpowered routers that simply do not have the muscle to do the job. Horsepower is an important consideration to bear in mind when choosing a router. Conversely, the other temptation is to buy one of the very large (and heavy) industrial routers. This is often unnecessary for the occasional woodworker; moreover, a heavy router can be awkward in many routing situations.

The more serious woodworker (and certainly the professional) has to consider owning two routers and perhaps a laminate trimmer. Besides a 1-h.p. router, which I in-

1-17: Some router bases have flat sections that accept C clamps. I prefer this type of auxiliary fence to the standard edge guide.

herited many years ago (and which now collects dust because its power is not great enough), I own a 1½-h.p. and a 2-¾-h.p. plunge router. I find that they complement each other beautifully; each one can do certain things better than the other. For general joinery, circle cutting, freehand routing, and rabbeting, I tend to use the lighter router. For real power routing, such as deep-slot cutting, deep mortising, or repetitive production work, I rely on the larger router. The plunge feature is an asset, but it is actually the increased horsepower that sets these two machines apart.

There are three additional considerations when choosing a router. The first is whether the router has the necessary housing to permit it to be clamped upside-down in a vise (**fig. 1-15**). I find this feature indispensable for quick setups and test cuts. Second, I prefer a router base that will accept C clamps for the purpose of clamping on an auxiliary fence (**fig. 1-17**). Third, and most important, the router must have the capacity to accept

bits with ½-inch (12-mm) shanks. Routers with only ¼-inch (6-mm) collets greatly limit the variety of bits the tool will accept. Routers with greater horsepower almost invariably accept bits with ½-inch shanks and are, therefore, more versatile machines.

I am tempted to say that I have never met a router that I did not like. This is almost true. The happy reality is that there are many excellent routers from which to choose. Talk to different woodworkers. Perhaps you can test their routers in real cutting situations. Try replacing a bit. Features tend to be comparable, but each router does have a different feel. Evaluate your own needs, ask questions, and look for a good value.

Before the router was developed, many woodworking tasks were far more laborious and time-consuming than they are today. And, as is the case with many mechanical innovations, the router came into being when its inventor, R. L. Carter, needed a better, faster way to get a job done. **Figure 1-18** shows an early model of the router.

THE FIRST ROUTER

By E.C. Benfield, *The Stanley Works*

During World War I, a young and ingenious patternmaker and mechanic extraordinary, R. L. Carter, owned and ran a pattern shop called The Carter Pattern Works of Syracuse, New York. He was a wood and metal patternmaker with a reputation for craftsmanship. He was also of an inventive nature.

Junius A. Yates, his nephew, now retired from Stanley Power Tools, a division of The Stanley Works, worked for Carter and before his retirement related these facts about his uncle and the opportune invention of the first portable electric hand shaper, which was to become the basis for the first portable electric router.

Carter had an order for a pattern and core box for a sectional boiler, one that was larger

1-18: An early Stanley router *(Stanley Works)*.

than average. The manufacturer had given his specifications for certain radii to be cut in the core box, all with matching edges. Somehow the radii on the 16 core boxes were overlooked and a casting was made. Much to Carter's dismay, it was not accepted. The radii had to be put on the core boxes. How to do it?

The conventional way was a long hard tedious carving-out process by hand, using a type of spoke shave. He couldn't see wasting time and money to do it all by hand. Carter went home, got an electric barber clipper, took the motor out, removed a worm gear from the motor, and proceeded to grind a radius into it. He couldn't guide this new-born cutter, however, in order to get an accurate cut. Putting a little more thought into it, he made two bevel guides, placed them on the shaft, hooked the cutter up in between the two guides and plugged in the motor.

Working in mahogany—end grain, cross grain, and with the grain—Carter and his nephew, Junius Yates, finished the job on the 16 core boxes in about two hours, a task that would ordinarily have taken seven men about three days to do.

The new hand shaper had served its purpose and it was put aside until about a year and a half later. Junius stopped in at a local cabinetmaker's shop and found the cabinetmaker laboriously cutting ornate curves on wooden walnut frames, the back used in popular sofas of that day. Junius went back to his shop to pick up the electric hand shaper, and showed the cabinetmaker how easily the same curves could be shaped with the still-unheard-of electric tool. The cabinetmaker wanted to buy it immediately. That made Junius take it back to Carter with a report on the tool's performance. Carter decided that it was time to use it in his own pattern shop.

About eight months later, after the Armistice, Ray Carter sold the pattern shop. He decided to experiment on the little shaper and he made his own pattern for the end shield. Then he had some models of the tool made in a machine shop. He sent a few to his former electric-disc-sander salesmen and soon the orders began to come in.

Junius, who was now employed by another company, worked nights and Saturdays. He set up a machine shop in his uncle's garage and the two men began to turn out 15 shapers a week. This went on for eight months. Suddenly, they realized that they had 750 on order. They in-

creased the garage space and hired back two of the former machine operators from the old pattern shop.

The hand shaper was immediately dubbed "the wonder tool," and 10 years later, there were over 100,000 in use. This tool really started the R. L. Carter Co. of Phoenix, New York, which eventually became the world's largest producer of direct motor-driven portable woodworking machines.

Shortly after this achievement, Ray Carter's agent in Philadelphia learned of a problem encountered by the Pennsylvania Railroad. A large number of the doors (on the wooden coaches of this period) required larger hinges than were originally installed. The agent came to Carter with the problem. Again, Carter conceived the idea that his hand shaper could be mounted in a base and be provided with a chuck to hold a cutting tool such as a rotary file. Using the same motor as the one in the hand shaper, he designed a chuck to take cutter bits with ¼-inch shanks and a base for the tool, essentially the same principle as that used in today's router. Thus the router, as yet unnamed, was created. It was delivered with three bits to the railroad, and received its name when Carter used it to make molding cuts, like those produced for years by the present Stanley hand routers.

He continued to improve the router and to develop new adaptations and uses for it. By 1929, when Stanley acquired Ray Carter's business, the router and the earlier shaper were well-known to cabinetmakers, home builders and other woodworkers. These tools were great time-savers, and moreover, they were well-adapted to precision workmanship. They were amazingly versatile.

The original Carter router was a small, low performance motor, maximum output ¼ h.p. It had a good control of depth. The frame was threaded 16 threads per inch; thus one revolution of the motor in the base resulted in change of depth of $\frac{1}{16}$ inch. Because of the comparatively large diameter of the motor casing, it was easy, and still is, to divide one rotation by quarters, eighths, sixteenths, or smaller increments to achieve depth setting within accuracy limits of .008 or better. No competitor has yet offered a more precise system.

2 The Bits

The majority of router bits available might appear as a blur to the uninitiated craftsman. There are literally dozens of bits from which to choose (Porter-Cable alone offers over 170), in different sizes and with different functions. Some come with ball-bearing pilots, others with steel pilots, and still others with no pilots at all. Should one buy steel or carbide bits? Is it better to purchase bits with ¼-inch or ½-inch (6-mm or 12-mm) shanks? Router bits can also be rather expensive, and many a new router owner has been shocked to find that a good selection can cost considerably more than the router itself.

An understanding of the bits and their respective functions need not be complicated. Bits, or cutters as they are sometimes called, can be grouped into several basic categories, each of which is discussed later in the chapter. First, it is important to gather some basic information.

SHANK SIZE

Router bits are available in shank sizes of ¼ inch (6 mm), ⅜ inch (9 mm), and ½ inch (12 mm), as shown in **figure 2-1**, but the greatest variety of bit types are available with ¼-inch (6-mm) shanks. In time, your bit collection will be fairly evenly divided between ¼-inch (6-mm) and ½-inch (12-mm) shank bits. I certainly recommend using ½-inch shanks when the bit is intended to make large cuts or when considerable lateral stress will be exerted on the bit, such as when you are edge jointing (see Chapter 6, *Secrets of Joinery*). A ½-inch shank bit costs just a few dollars more but is more rigid, is virtually impossible to break, and chatters less. Chatter is the manifestation of vibration and results in an unsmooth cut (**fig. 2-2**).

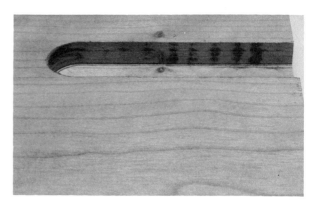

2-2: Chatter, a result of vibration, produces an unsmooth cut. Half-inch shank bits chatter less than ¼-inch (6-mm) shank bits.

2-1: From left to right: ½-inch (12-mm), ⅜-inch (9-mm), ¼-inch (6-mm) shank bits. Notice that the middle bit has a ball-bearing pilot mounted above its cutters.

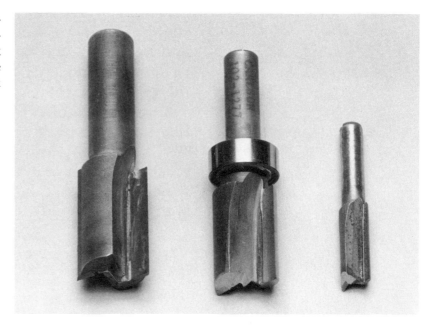

FLUTES

The number and types of cutting edges, or flutes, help to determine the cutting characteristics of a router bit (**fig. 2-3**). Small bits, ⅛ inch (3 mm) or less, are usually single fluted. Single-fluted bits are a little less expensive and they cut quickly because of the increased chip room, but I recommend using only double-fluted bits. Double-fluted bits provide twice as many cutting strokes per revolution, thereby producing a smoother cut. Single-fluted bits are used more in industrial routing, where speed is critical and where the cut will not be seen. A single-fluted bit is not a balanced bit and can strain the router's bearings. Bits with three and four flutes are also available, but are less commonly used because they cut more slowly and are more expensive.

Spiral-fluted bits are used principally for deep slotting and mortising (**fig. 2-4**). To this end, they are designed to pull chips up and out of the work. This chip removal results in

2-4: Spiral-fluted bits are used principally for deep slotting and mortising, and are designed to pull chips up and out of the work.

less vibration than if you were to use a straight cutting bit. Spiral-fluted bits with large diameters tend also to have long cutting lengths and may not be suitable for shallow work. Therefore, bear this in mind whenever you order these bits.

2-3: The number and types of cutting edges (flutes) help to determine the cutting characteristics of a bit. Left to right: a double-fluted, straight cutting carbide bit; a double-fluted, carbide coving bit; a four-wing carbide slotting cutter; a single-fluted straight cutting bit; a spiral-fluted bit.

CARBIDE VS. HIGH-SPEED STEEL

Router bits, which typically weigh between 3 and 4 ounces, are available in either high-speed steel or carbide tipped (**fig. 2-5**). The high-speed steel bits are made from the same high-speed tool steel used for lathe bits. Carbide-tipped bits (very few bits are made completely with solid carbide) have cutting edges of tungsten carbide brazed to the body of the bit (**fig. 2-6**). Because of its extreme hardness, a carbide-tipped cutter will make a finer cut over a longer period of time than its steel counterpart. As a corollary, it will put less strain on the router. Carbide bits are more resistant to abrasion and are ideally suited for cuts on tough, abrasive materials, such as plastic laminates, plywood, and particleboard. On the other hand, they are much more expensive than steel bits (often three times as high) and must usually be "sent out" to be sharpened. Should a tip chip or

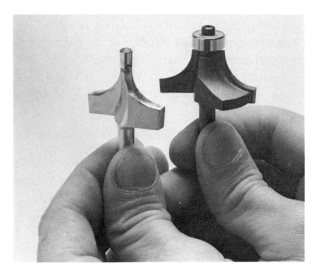

2-5: At left, a high-speed steel rounding-over bit. At right, the same bit in carbide. The pilot of the high-speed steel bit is molded of the same piece. Note that the carbide has a ball-bearing pilot.

2-6: Carbide-tipped bits have cutting edges of tungsten carbide brazed to the body of the bit. This bit has a ball-bearing pilot, indicating that the bit is intended to ride along the edge, rather than through, the body of a piece of wood.

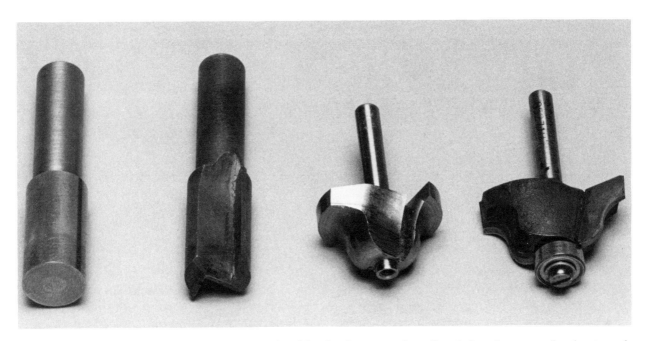

2-7: Left to right: a ⅝-inch (15.8-mm) bit blank; the same bit after it has been cut back, tipped with carbide, and reground; a Roman-ogee blank; the same bit in a finished state.

break (and occasionally they do), it is very expensive to have it rebrazed.

I almost categorically recommend the purchase of carbide bits, however. They cut so much more easily and with so much better results that, in the long run, they are a much better investment. Steel bits have more of a tendency to burn the wood, to dull more quickly, and are practically useless in any kind of production situation. In addition,

steel bits are not available with ball-bearing pilots, which is a major drawback.

Steel bits do have their place, however. They do make a better initial cut, since their edges have a finer, factory-honed edge. They also permit the newer woodworker to have a fuller collection of bits (initially) at a moderate price. Certainly bits that are not going to be used frequently can be purchased in steel. You get what you pay for.

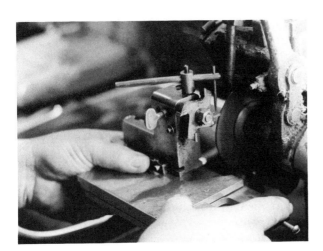

2-8 (left): Grinding a slotting cutter to finished dimension. *(Photo courtesy of the Fred M. Velepec Co).* **2-9 (below):** A single-fluted carbide bit. *(Photo courtesy of Ekstrom Carlson)*

PILOTS

At the base of many router bits you will find what is called a pilot (**fig. 2-5**). The pilot rides alongside the wood, thereby controlling the amount that the bit will cut into the edge. Any time you see a bit with a pilot, you know that it is intended for cutting along the edge of a board rather than within the body of the wood. On steel bits the pilots are usually a molded part of the bit, although a number of manufacturers offer a screw-in pilot that permits different size pilots to be interchanged (**fig. 2-10**). Carbide bits are fitted with ball-bearing pilots (**fig. 2-6**), which are often protected by a dust shield (**fig. 2-11**). Ball-bearing pilots provide a much smoother ride and will not burn the wood. Excessive pressure on a pilot will, however (particularly on soft woods), sometimes leave a mark, so do not bear down more than you have to (**fig. 2-12**).

Most ball-bearing pilots can be removed by loosening either its machine or its Allen screw (**fig. 2-13**). By interchanging pilots the profile of a bit can be changed. The rounding-over bit shown in **figure 2-14** becomes a beading bit when a smaller pilot is used.

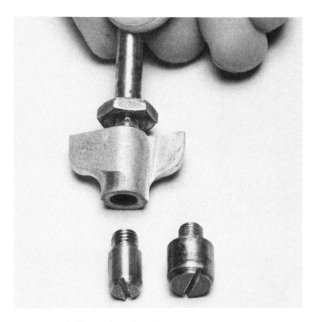
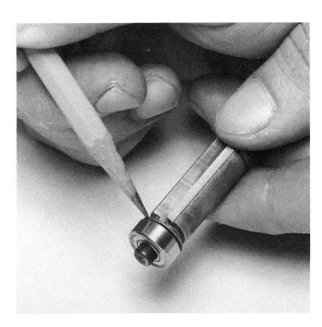

2-10 (left): This bit has a screw-in pilot, which permits different size pilots to be interchanged.
2-11 (right): Some bits, like this flush-trim bit, have a dust shield that helps keep the bearings between the bit and the pilot free of dirt, pitch, or in this case, contact cement.

2-12: The piece at left was scored by excessive pressure on the ball-bearing pilot. The piece at right was burned by excessive pressure on a high-speed steel pilot.

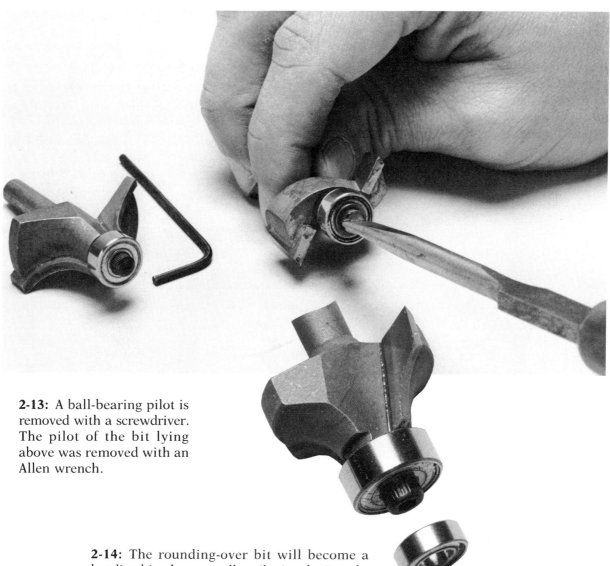

2-13: A ball-bearing pilot is removed with a screwdriver. The pilot of the bit lying above was removed with an Allen wrench.

2-14: The rounding-over bit will become a beading bit when a smaller pilot is substituted.

UNIVERSAL ARBORS

Although the arbor (shank) and body of a bit are usually molded of a piece, it is possible to purchase individual arbors onto which ball bearings and any number of different cutters are fitted, including carbide-tipped cutters (**fig. 2-15**). This universal arbor system offers a less expensive means of owning a number of different profiles, although few woodworkers seem to take advantage of it, preferring instead to use one-piece bits. This is probably due to ineffective marketing on the part of the manufacturers, coupled with the fact that the bits are not always available in carbide. The universal arbor system is also used to mount slotting cutters.

2-15: Interchangeable cutters fit on this universal arbor made by Bosch.

MAINTENANCE, SHARPENING, AND STORAGE

The most important factor in bit maintenance is keeping the bits clean. Pitch and dirt build up on the cutting edges, which in turn cause excessive heat buildup during the cutting process. Pine is probably the worst culprit. Its acid residue will actually attack the cobalt (the cementing agent in the carbide) and pit the carbide (**fig. 2-16**). Unclean bits dull quickly and cut poorly. The easiest and fastest method for cleaning bits is to use an oven cleaner. Spray or wipe on the cleaner and let it sit for a few minutes (**fig. 2-17**). A wet sponge or rag will then remove the pitch (**fig. 2-18**).

2-16, 2-17, and **2-18:** An unclean bit at left. Middle: The bit is covered with oven cleaner and allowed to sit for a few minutes. Right: After the bit is sponged with water and dried, it shines like new. I have tried many methods for cleaning bits and sawblades and, by far, this is the best and fastest way.

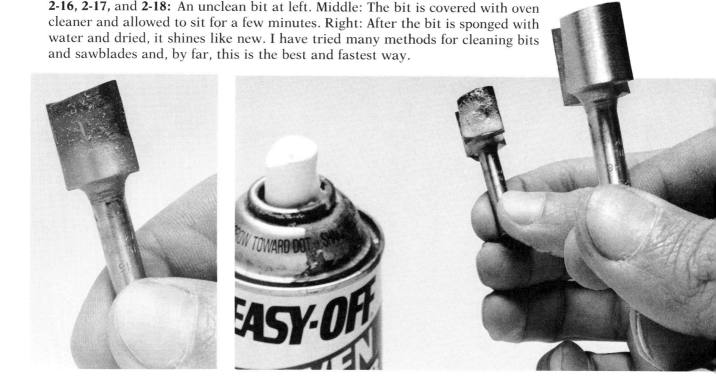

Ball-bearing pilots should spin freely; when they do not, they must be cleaned. Gumming of the ball bearing occurs most frequently on flush-trimming bits, which, when trimming plastic laminates, become clogged with contact cement (**fig. 2-19**).

The bearing and dust shield should be removed from the bit and cleaned with lacquer thinner (**fig. 2-20**). If the bearing cannot be removed, immerse the whole bit in a jar of lacquer thinner. Afterward, lubricate with a light machine oil. Pilots, by the way, will occasionally freeze (an indication that they are broken) and should simply be replaced.

High-speed steel bits, which should be kept sharp, can be honed lightly on a fine stone such as a soft Arkansas (**fig. 2-21**). Concentrate on honing the flat, inner sides of the bit. The beveled cutting edges can also be "touched" lightly with a round, tapered stone (**fig. 2-22**). If you are bold and have a few old bits with which to practice, try

2-19 (right): Flush-trimming bits will gum up with contact cement when used for trimming plastic laminates.

2-20: To clean a flush-trimming bit, soak the bit in lacquer thinner. Then, remove the pilot and dust shield and clean thoroughly. Reassemble and lubricate with a light machine oil.

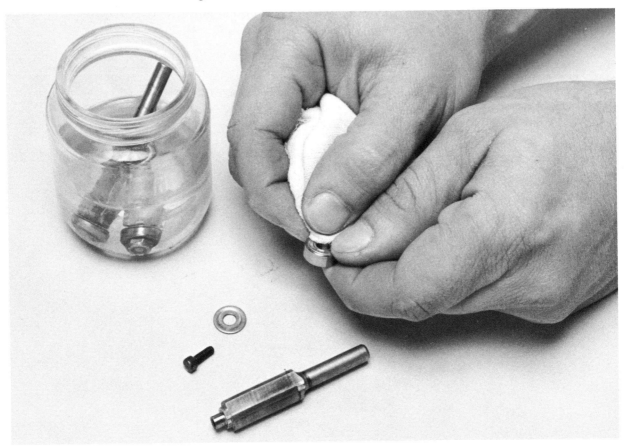

sharpening on a grinding wheel (**fig. 2-23**). Use a 60-grit wheel and grind on the flat side of the wheel. Again, work only the flat, inner sides of the bit. Too much grinding will change the bit's diameter. Work both sides of the bit evenly, or only one flute will cut.

Usually, carbide bits must be sent out to a professional sharpening service. Again, with carbide bits, the emphasis should be on keeping them clean. By doing so you will be amazed at how long they will stay sharp. **Figure 2-24** shows one way of storing your bits, to keep them from cluttering your workplace or picking up loose dirt.

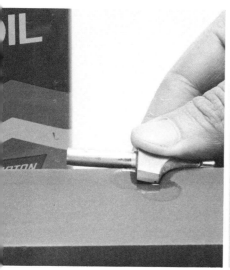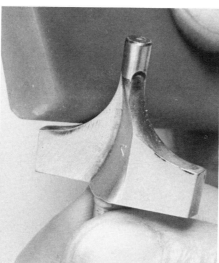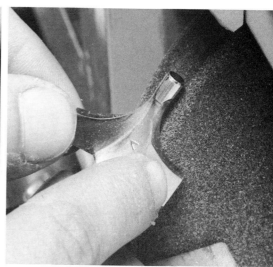

2-21 (left): High-speed steel bits can be honed lightly on a fine stone. Work the flat, inner portion of the bit. **2-22 (middle):** The beveled cutting edges can also be honed with a round, tapered stone. The one shown here is a soft Arkansas stone. **2-23 (right):** When sharpening a bit on the grinding wheel, hold the flat, inner faces of the bit against the flat side of the stone. Use a 60-grit wheel and do not try to remove too much metal.

2-24: Typical bit storage. As shown, I have one block for ¼-inch (6-mm) bits and one for ½-inch (12-mm) shank bits.

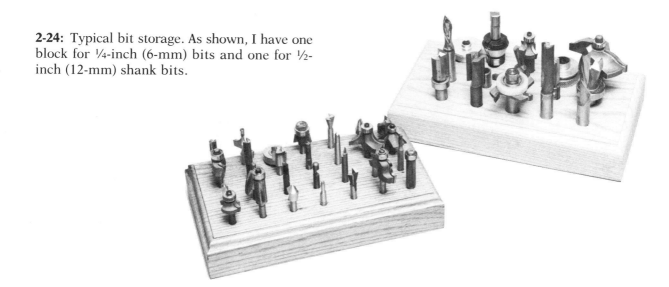

TYPES OF BITS

Some router bits can be grouped according to function, while others do not fit neatly into a given category. Let's take a look at some of the more common groupings.

Straight Bits

The most extensively used of all router bits, straight bits are used to make grooves, rabbets, dadoes, mortises, and tenons. Straight bits are also used in a multitude of other operations, including edge-to-edge jointing, hinge mortising, circle cutting, and so on (**fig. 2-25**). A straight bit, when viewed from above, is circular, which explains the rounded end it leaves on a stopped groove. Different manufacturers offer the bits in different widths. The smallest size available is ¹⁄₁₆ inch (1.5 mm). Other common sizes available include ⅛ inch, ¼ inch, ⁵⁄₁₆ inch, ⅜ inch, ½ inch, ⅝ inch, ¾ inch, ¹³⁄₁₆ inch, ⅞ inch, and 1 inch (3, 6, 7.9, 9, 12, 15, 19, 20.6, 22, and 25 mm). Lengths will vary also. The length of the cutting edge on a ¼-inch (6-mm) bit, for example, can vary from ⁵⁄₁₆ inch to 1⅛ inch (7.9 to 27 mm). Some bits with ½-inch (12-mm) shanks can be purchased in lengths of up to 2 inches (50.8 mm).

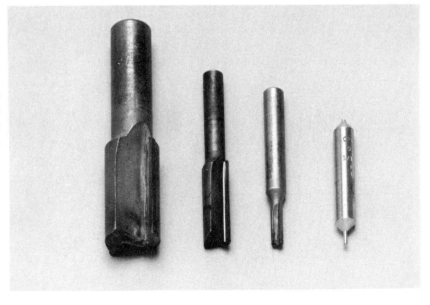

2-25: From left to right, a ¾-inch (19-mm) carbide bit with ½-inch (12-mm) shank; a ⅜-inch (9-mm) carbide bit with ¼-inch (6-mm) shank; a ³⁄₁₆-inch (4.7-mm) carbide bit with ¼-inch (6-mm) shank; a ¹⁄₁₆-inch (1.5-mm) high-speed steel bit with ¼-inch (6-mm) shank.

Edge-Forming Bits

Of all the router bits, the decorative edge-forming bits raise the most eyebrows, since they can transform the plain edge of a board into a classic shape with craftsmanlike precision in just a few moments. It is extraordinary to see what a difference a stopped chamfer can make on the post of a trestle table or how a Roman ogee edge can add elegance to the top of a nightstand. Edge-forming bits are also used to make moldings.

The Roman Ogee. I will never forget the student of mine who announced that he was ready for a "Roman orgy." A truly classic shape, the Roman ogee edge can be incorporated into many styles of furniture. **Figures 2-26** and **2-27** show applications of this bit. Some router bit manufacturers also offer a simple ogee bit.

2-27: A Roman ogee was used to make this elegant detail for a baseboard molding.

The Chamfer Bit. The chamfer bit (**fig. 2-28**) is too often overlooked by the amateur woodworker, who often opts for a more dramatic-looking profile. Originally used to take

2-26: The Roman ogee bit is used to cut a shape for a table top.

2-28: A stopped chamfer on a trestle post.

the sharpness off the corners of boards (beveling), the chamfer produces a delicate touch on even the most rustic of furniture.

The Rounding-Over Bit. Sometimes referred to as a "quarter-round," this bit produces a beautiful rounded edge (**fig. 2-29**). The cutter can be used at different depths to produce varying results. When it is dropped a little past maximum radius, a thumbnail molding, such as the one shown in **figure 2-30,** is produced. When used with a matched cove cutter, a rule joint can be made (**fig. 5-5**). The rule joint is a choice method for connecting a dropleaf to a table. Use a smaller pilot, and the rounding-over bit becomes a beading bit (**fig. 2-14**).

2-29 (left): I used a rounding-over bit to produce a soft edge on this bannister rail. **2-30 (right):** The bottom molding on this Shaker reproduction was made with a rounding-over bit dropped just past maximum radius. Called a "thumbnail" molding, this detail is echoed on the drawer front.

The Beading Bit. The beading bit produces a fine detail for period furniture (**fig. 2-31**). The rounded edge formed by a beading bit is similar to the radius produced by a rounding-over bit, except that the beading is punctuated by two shoulders. For examples, this beading is commonly used as a decorative detail on door and cabinet frames, table tops, and aprons.

2-31: The beading bit produces delicate moldings such as these.

The Cove Bit. The cove bit, as shown in **figure 2-32**, is used for a wide variety of tasks. It may be used to make moldings, rule joints, and finger-grip handles for doors and drawers (**fig. 2-33**).

2-32: The cove bit was used here to make a delicate edge on this zebrawood box.

2-33: The finger grip for the back of this door was made with a cove bit.

2-34: This photo shows a ⅜-inch (9 mm) rabbeting bit. The ⅜ inch (9 mm) indicates how far the bit cuts in, not down.

The Rabbeting Bit. The rabbeting bit is essentially a straight cutting bit with a pilot added. The distance from the pilot to the outer edge of the cutter determines the amount to be rabbeted, or cut away (**fig. 2-34**). The depth of the rabbet is determined by the position of the router base. Rabbeting bits can form rabbets from ¼ inch to ¾ inch (6 to 9 mm), but rabbets of any size can also be made by using a straight cutting bit in conjunction with an edge guide.

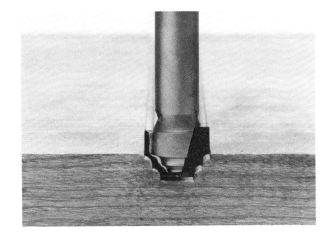

2-35: A typical groove-forming bit, designed to cut within the body of a piece of wood.

Groove-Forming Bits

Groove-forming bits are similar to decorative edge bits, except that they are designed to cut within the body of wood rather than along its edge. As such, they do not have pilots, so you will need another system for guiding the router (**fig. 2-35**). See Chapter 7, *Pattern and Template Routing*. Groove-forming bits are often used on doors and drawer

2-36: The V-groove bit is used to create the effect of V-groove paneling on this door.

faces to simulate the effect of frame and panel construction.

The V-Groove Bit.
The V-groove bit is an interior decorative bit that is often used to simulate the effect of a carver's parting tool. It can also be used in lettering, where letter templates are used, and to simulate V-groove paneling (**fig. 2-36**).

The Core-Box Bit.
The core-box bit, which produces a semicircular groove, is typically available in widths from ¼ inch to ¾ inch (6 to 19 mm). It is especially effective in fluting flat surfaces (**fig. 2-37**). The core-box bit is also used for making drainboard fluting on carving boards and for making finger grips for drawers and doors.

The Veining Bit.
The veining bit and its close cousin, the round-nose bit, are used for general decorative line work and fluting.

The Slotting Cutter.
The slotting cutter, as shown in **figure 2-38**, is a general grooving bit, particularly useful for making spline grooves. It is available in kerfs of ⅟₁₆ inch (1.5 mm) to ¼ inch (6 mm).

2-37: The core-box bit.

2-38: The slotting cutter.

Laminate-Trimming Bits

Laminate-trimming bits are most often used to trim plastic laminates (such as Formica) flush with the wood to which they have been glued. One type of laminate-trimming bit, the flush-trimming bit, has a ball-bearing pilot and carbide-tipped edges (**fig. 2-11**). Other trimming bits will not only flush the overhanging laminate, but will also finish it with a beveled edge, the angle varying from 7–22 degrees. The flush-trimming bit is actually one of the most useful bits in the woodworker's arsenal. You will use it to flush contiguous pieces of wood and in repetitive pattern cutting.

Other Common Bits

The Hinge-Mortising Bit. This bit is designed for cutting mortises for hinges because it cuts fast and is capable of plunging. (Plunging is the lowering of a bit into a piece of wood.) Straight cutting bits, however, can

2-39: The dovetail bit.

also plunge and are often substituted for hinge-mortising bits. Since the amateur or occasional woodworker would probably use the multi-purpose straight cutting bit for these tasks rather than purchase both bits, the hinge-mortising bit is usually reserved for professional woodworkers.

The Dovetail Bit. When used in combination with dovetail templates, this bit, as shown in **figure 2-39,** can produce the dovetail joint, which many feel is one of the finest joints in cabinetry. It can also be used to make slotted dovetails (**fig. 10-27**).

Custom Bits

Sometimes the solution to a woodworking problem is arrived at with a specifically designed router bit. Many independent bit manufacturers will customize bits, and they can make just about anything you can imagine (**fig. 2-40**). Put your idea into a sketch. Let them tell you if the design is feasible and what it will cost. An excellent example of how a custom bit is used to solve a problem is found in **figure 10-22.**

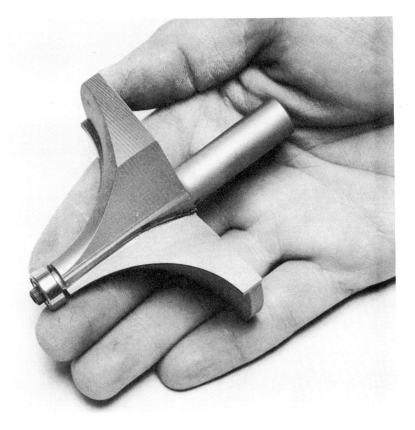

2-40: Either I have a very small hand or this is a very large bit! Actually, this is a jumbo, custom-made rounding-over bit with a 1⅛-inch (28-mm) radius, made by the Fred M. Velepec Co.

RECOMMENDED BASIC KIT OF ROUTER BITS

When reading through a list of cutters, such as the one presented on preceding pages, it is very easy to fall into the basic buyer's trap: "Oh, I will need this one, and I will definitely need that one," and so on. It is not necessary to own every cutter; indeed, it would cost a small fortune. Many cutters can perform multiple roles. The ¾-inch (19-mm) straight cutting bit, for example, can be used not only to make ¾-inch (19-mm) and wider dadoes, but it can also make rabbets of any size. The same cutter can also cut a hinge mortise and even be used to make circles.

If you want to make dovetails you will need a dovetail bit. Generally speaking, however, purchase your router bits as you need them.

Below is a list of what I consider to be a good basic kit. Buy all carbide-tipped bits if your budget permits.

- ¼-inch (6-mm) straight cutting bit (carbide)
- ½-inch (12-mm) straight cutting bit (carbide), with ½-inch (12-mm) shank and the longest cutting edge available
- ¾-inch (19-mm) straight cutting bit (carbide)
- Roman ogee edge (high-speed steel or carbide)
- ½-inch (12-mm) rounding-over bit (high-speed steel or carbide)
- chamfer bit (carbide or high-speed steel)
- ⅜-inch (9-mm) rabbeting bit (carbide with ball-bearing pilot)
- flush-trimming bit—1 inch (25 mm) in length (carbide only)

3 Cutting Methods

The router is typically used in one of two ways: either it is guided along or through a piece of wood, or the wood itself is fed along a mounted router. In either case, there are certain universal rules that govern proper router usage. These rules are based largely on the physics involved, which will help you determine the proper direction and rate of feed. This knowledge greatly affects the accuracy and quality of your work, as well as your own safety.

With the wood held stationary, you can use various guides for the router or you can operate it freehand.

With the router stationary, either clamped upside down or mounted upside down in a router table, you move the wood over the router bit.

Each method has its advantages.

MOVING THE ROUTER

In moving the router, you have several options for controlling the cut.

Using a Standard Edge Guide

A standard piece of equipment on all routers is an adjustable fence that is designed to ride along the edge of a board (**fig. 3-1**). You will use the edge guide (fence) perhaps more than any other attachment to the router. Edge guides are connected to the router base with guide rods. The guide rods are held in place either by setscrews or thumbscrews. These provide the gross adjustment for the guide. Some edge guides have, in addition, a vernier adjustment that allows fine tuning (**fig. 3-2**). Edge guides that lack a fine-tuning component can be exasperating to work with, because tightening the setscrews or thumbscrews tends to shift the fence out of position.

Most edge guides come with predrilled holes so that an auxiliary wooden fence can be screwed to it. This added fence provides a larger bearing surface and lends greater routing stability. If your guide is not predrilled, drill the holes yourself. Wooden fences, by the way, are best made from a hardwood such as maple or cherry, which will polish naturally from continued use.

3-1: A standard edge guide attached to a router with guide rods. This particular edge guide is not outfitted with a vernier adjustment.

You can make a wedge-shaped edge guide, as shown in **figure 3-3,** to use when routing round edges.

A simple alternative to the standard edge guide, and one that I use a lot, is a wooden fence secured to the router base with two C clamps (**fig. 3-4**). It is quickly and easily adjusted by loosening one clamp and tapping the fence in or out. This system is not possible on some routers, owing to limited clamping space. But if your router can accommodate C clamps, this may be the best edge guide yet, since it is less time consuming than adjusting a standard edge guide.

Using a Piloted Bit

The pilot of a bit rides along the edge of a board, thereby controlling how far the bit will cut in. When cutting along the edge of a straight board, I often use an edge guide in conjunction with a piloted bit (**fig. 3-5**). The edge guide prevents the pilot from leaving any marks on the wood and gives the router that much more stability. As shown

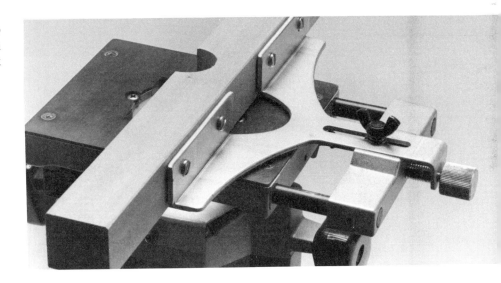

3-2: This router has a knob for fine tuning the position of the edge guide. A wing nut locks the guide in place.

3-3 (below): A wedge-shaped edge guide for routing round edges has been added to this standard edge guide, and is held in place with screws. **3-4 (below right):** A fence clamped to the router base with C clamps may be the best edge-guide system of all.

3-5: Here, an edge guide is used in conjunction with a piloted bit. Although the bit does not require it, the addition of the edge guide will give you greater control over the cut. It will also prevent the pilot from burning the wood.

in **figure 3-5,** the fence is notched so that as much or little of the bit can be exposed as desired. A piloted bit will have a tendency to "turn the corner" when it reaches the end of a board no matter how careful you are. Attaching an auxiliary fence to the edge guide can prevent this.

Using a Fence

Clamping a fence, or straightedge, to the work surface, as shown in **figure 3-6,** is one of the most common ways of guiding the router through the body of a piece of wood. Carefully measure the distance between the outside cutting edge of the bit and the outside edge of the router base in order to po-

sition the straightedge. Fences should be at least 2½ inches (60 mm) wide so as not to deflect during a cut.

An excellent variation of the fence is the T-square straightedge (**fig. 3-7**). It squares the straightedge to the edge of the board, and the groove in the jig allows for quick alignment. (Using the groove for alignment purposes presupposes a different jig for each size router bit.)

Using a Trim Guide

If there is one section in this book to underline in red, this is it. Several years ago, a fellow cabinetmaker showed me a jig for which he had no name. I dubbed it the "trim

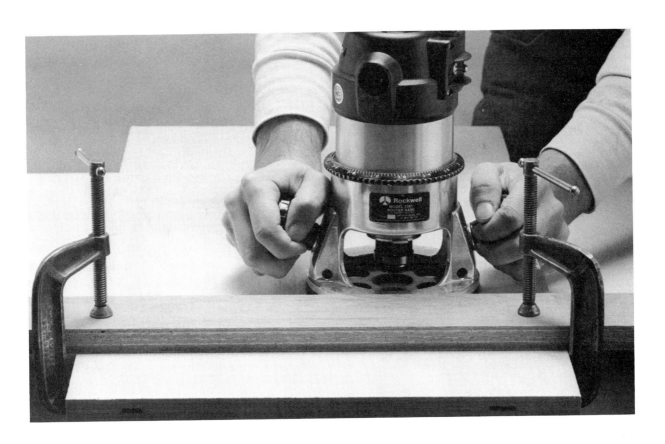

3-6: One common way to guide a router is along a straightedge clamped to the workpiece.

3-7: An excellent fence for 90-degree routing is the T-square straightedge. Remember, if you use the groove in the jig to align your cut, you will need a different jig for each size bit.

3-8: Nailing the trim guide together.

guide." You will find dozens of applications for it. Before I tell you how to use one, let me tell you how to make one. Begin by nailing and gluing a piece of plywood ¼ inch × 6 inches × 36 inches (6 mm × 15.2 cm × 91.4 cm) to a perfectly straight piece of wood (solid or plywood), ¾ inch × 3 inches × 36 inches (19 mm × 7.6 cm × 91.4 cm) (**fig. 3-8**). Insert a straight cutting bit into your router and, guiding the router against the ¾-inch (19-mm) piece, make a straight cut (**fig. 3-9**). What you are left with is a trim guide.

Cutting a piece of wood on an angle provides a good example of how the guide can be used (**fig. 3-10**). First, mark the angle and, with a saber saw, cut along the outside of this line. Then, place the edge of your guide directly on the line. The router is now automatically positioned, avoiding any measuring or clamping of fences. Understand that

3-9: Completing the trim guide by cutting the ¼-inch (6-mm) wood with a straight cutting bit. Remember, this trim guide will always be used with the same size router bit.

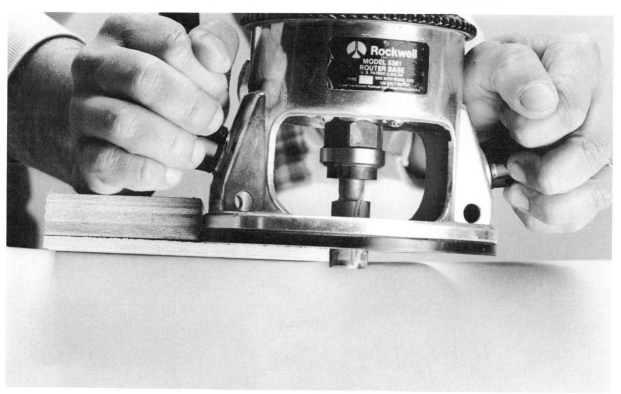

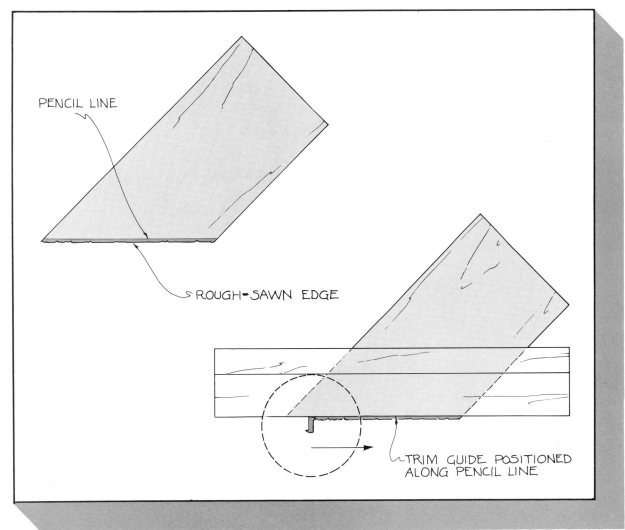

PENCIL LINE

ROUGH-SAWN EDGE

TRIM GUIDE POSITIONED
ALONG PENCIL LINE

3-10: Cutting a piece of wood on an angle provides a good example of how a trim guide can be used. A saw removes the bulk of the wood. The router, guided against the fence of the trim guide, cuts the angle straight and true.

the bit will always cut exactly along the edge of the jig, as long as the same size bit is used as was originally used to make the jig.

Although the router was not designed for ripping or crosscutting, the trim guide does provide you with a means of trimming boards to length and width (**fig. 3-12**). Any significant amounts of wood should be removed with a saber or circular saw. **Figure 3-11** shows two boards being trimmed to the exact same length with a router.

I sometimes use the trim guide in lieu of hand planing when fitting doors. It is also very useful on installations when counter-tops, for example, have to be scribed quickly to out-of-square walls. I have trim guides from 16 inches to 5 feet long (40.6 cm to 4.6 m) (**fig. 3-13**).

In addition, the jig can also be used to make dadoes by simply regulating the depth of cut. Remember, you will have to have a different size trim guide for each different cutter. Clearly mark on each of your guides which size bit was used to make it.

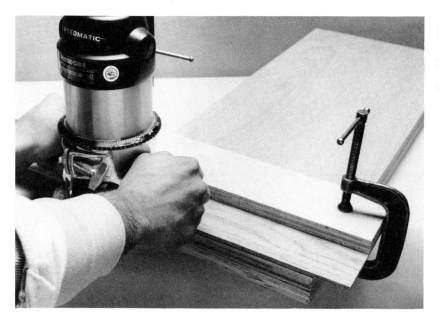

3-11: With the edge of the trim guide placed exactly along the line of cut, two boards are trimmed to the exact same length.

3-12: To trim a board, whether it is straight or on an angle, place the guide (top) on your pencil line and rout. In this case, no measuring is necessary.

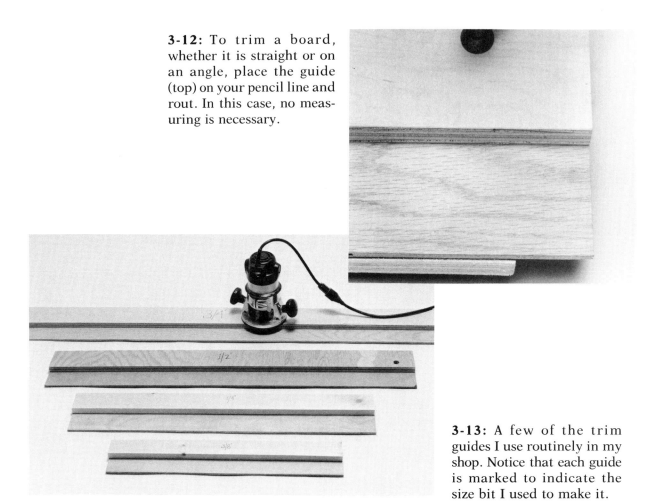

3-13: A few of the trim guides I use routinely in my shop. Notice that each guide is marked to indicate the size bit I used to make it.

Using a Template and Template Guide

Template guides are used for cutting irregular shapes and patterns. The template guide (also called a *bushing guide)* is a tubular piece of steel that fits around the bit and is secured to the router base. In some cases, the guide is actually screwed into the base and in others is held in place by a locking nut (**fig. 3-14**). The template guide rides against a premade template, or shaped pattern (**fig. 3-15**). In this way, the shape of the template is transferred to the wood.

Template guides come in different sizes with varying inside and outside dimensions and in different lengths. In cutting the pattern or template to be used, you must first measure the distance from the edge of the router bit to the outside of the bushing guide (**fig. 3-16**). The template must then be cut smaller *or* larger than the final desired shape by whatever this distance is, depending on the shape to be cut. For instance, if you were to make a shallow U-shaped template and

3-14: This template guide is secured to the router sub base with a locking ring such as the one shown in the foreground. Some template guides are held in position with machine screws. Each router manufactuer makes its own template guides.

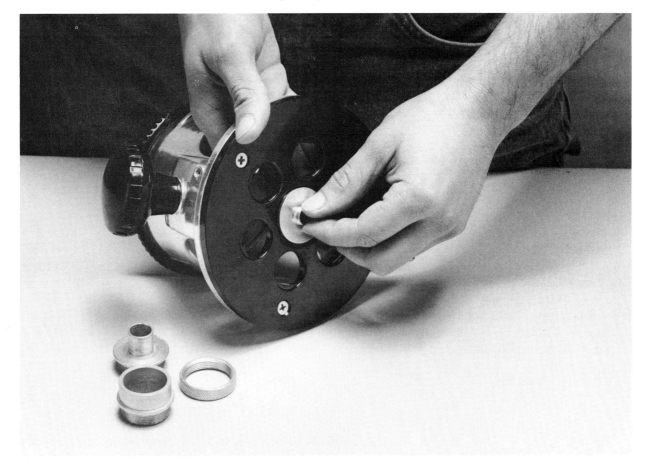

3-15: The template guide rides against a premade template. In this way, the shape of the template is transferred to the wood.

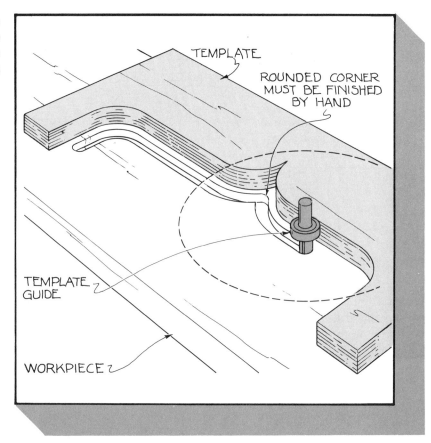

the template guide were to follow the *inner* curve of the U, your finished workpiece would be slightly smaller than the template. If, however, you constructed a U-shape template in which the template guide would follow the *outer* curve of the U, your workpiece would be larger than the template. Adjust the size of the template to compensate for the desired size of the workpiece.

Template guides are also used in decorative line work, dovetailing, and butt-hinge cutting. For more on this, see Chapter 7, *Pattern and Template Routing.*

Freehand Routing

Guiding the router with your hands can be done, but this requires skill and practice. This technique is used in lettering, in background removals for carvings, and sometimes in hinge mortising. For more on this, see Chapter 11, *Freehand Routing.*

3-16: Measuring the distance from the edge of the router bit to the outside of the bushing guide. Templates have to be made smaller or larger than the final desired shape because of this dimension. (The template shown in **fig. 3-15** is larger than the final shape of the workpiece.)

MOVING THE WOOD

If the router can be moved along the wood, then the wood can also be moved onto a stationary, inverted router. In many cases, particularly if the wood you are working with is small, it is easier to move the wood. Imagine cutting a tenon on narrow stock (**fig. 3-17**). Cutting a tenon with a moving router is awkward and time consuming, since you have to unclamp the wood, turn it over, and reclamp it after every cut. **Figure 3-18** shows how much easier it is to cut the tenon on an inverted router.

There are two basic systems to be employed in inverting the router.

Clamping the Router Upside Down

Clamp the router upside down in a woodworker's vise (**fig. 3-19**). Not all routers can be clamped this way; however, and your router must have mounting "ears" or flanges for this purpose (**fig. 1-15**). Clamping a router not meant to be clamped could distort the

3-17 (left): Cutting a tenon with a moving router is awkward and time consuming. **3-18 (below):** On an inverted router, tenon cutting is easier.

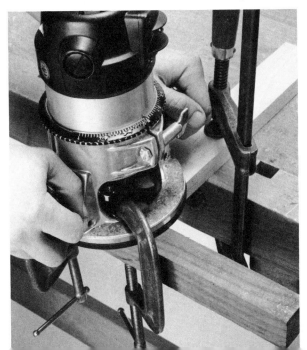

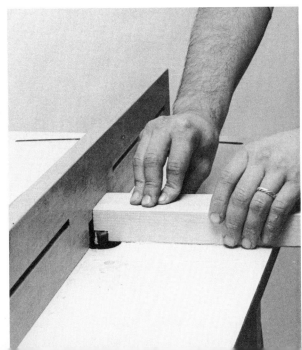

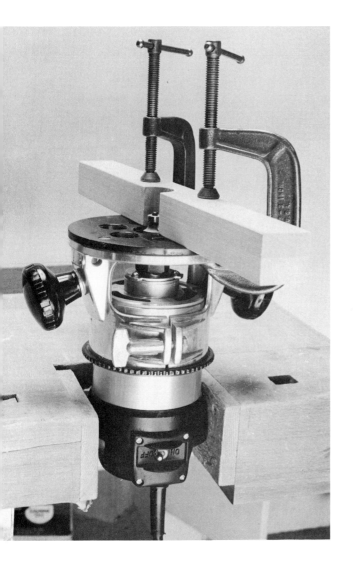

case and actually damage its bearings and armature.

I use my 1½ h.p. router in this clamped position all the time. Even when I plan to move the router along the wood, I first lock it in the vise to set the depth of cut and position the fence. While it is in this position, I make a test cut to check the accuracy of the settings. I then remove the router and make my cuts in the wood.

Using the Router Table

The other system involves the use of a router table (see Chapter 5, *Edge Routing*). By inverting the router and attaching it to a table, you have created a broad surface to support your work. And, in effect, you will have made a small shaper (**fig. 3-18**).

3-19: One way to invert a router is to clamp it upside down in a woodworker's vise. Unless your router has the necessary flanges to do this, you run the risk of distorting the case.

THE PHYSICS OF ROUTING

Understanding the physics of routing is not only interesting, but also practical. It helps you to understand many routing phenomena, such as kickback, grain tearout, and "wandering." Perhaps most important, it gives you general rules for proper direction and rate of feed. For years, I was intrigued by the question of whether it mattered which way a router was moved along a fence when cutting within the body of a piece of wood (**fig. 3-20**). "Left to right" and "right to left" are relative to where one is standing and certainly the router cannot make that distinction. Yet there is, in this case, a correct direction for moving the router.

Let us begin by examining a router cut made along the edge of a board. When viewed from above, a router bit spins in a

3-20: When routing against a fence within the body of a board, it does matter which way (left or right) the router is moved.

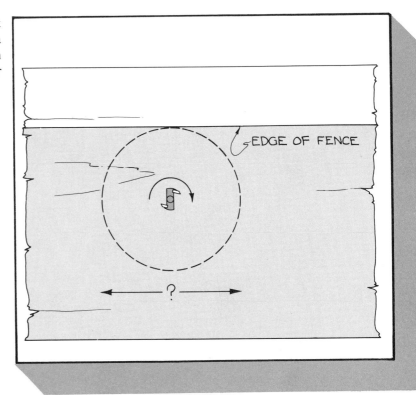

EDGE OF FENCE

?

clockwise direction (**fig. 3-21**). This spinning bit will exert a force on the wood with which it comes in contact. The wood, in return, will exert a force back against the bit. Imagine if you were to slowly spin a router bit by hand against a firmly clamped piece of wood (**fig. 3-22**). The router itself would be pushed away from the wood. Certainly, anyone who

has ever accidentally turned on a router while the bit was in contact with the wood has experienced this phenomenon. This is known as kickback. Kickback will always occur in the direction opposite from the spin of the cutter. (In other words, if the bit is spinning clockwise, the router will be pushed back counterclockwise.)

3-21: When viewed from above, a router bit spins in a clockwise direction.

3-22: If you were to spin a router bit slowly by hand, the router would be pushed away from the wood.

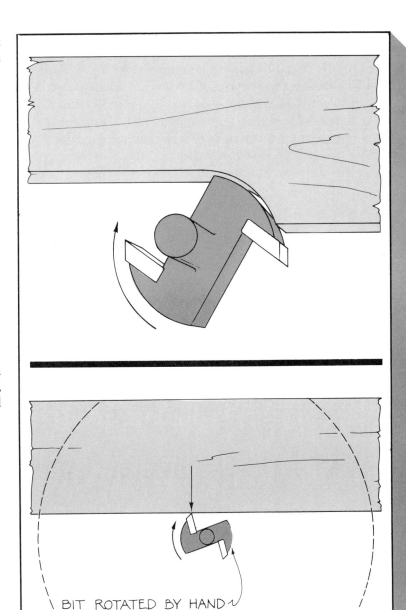

BIT ROTATED BY HAND

All of this would suggest that when routing along the edge of a board, the router would be constantly kicking back. This is not the case, however. In fact, if the router is fed in the proper direction, the spinning bit will actually propel the router against the wood. To understand this concept of "proper" direction one must first see why left-to-right and right-to-left routing result in two distinctly different types of cutting action. **Figure 3-23** shows the type of cut produced when routing from left to right. When I say "left to right" or "right to left," it is assumed that you are facing the wood and that the router is between you and the wood.

Look at **figure 3-23.** Notice that the cutting edge at A is meeting no resistance, whereas the leading cutting edge (B, in this case) is encountering considerable resistance. The wood pushes back against cutting edge B, and the router is pulled into the workpiece. In addition, in routing from left to right, the bit tends to "scoop" the wood in what is essentially a ripping action.

Right-to-left edge routing results in a very different type of cutting action. Think of the cutting edges in **figure 3-24** as small chisels and you can see that cutting edge A is cutting almost directly across the grain. If you have worked with chisels, you know that cutting against the grain meets with considerably more resistance than cutting with the grain, although a cleaner cut is usually achieved against the grain. Notice also that the resistance at point A will have a tendency to push the router away from the wood.

Now, let us go back to the question at the beginning of this section. Does it matter

3-23: In left-to-right edge routing, cutting edge A meets considerably less resistance than cutting edge B. The resistance met at point B tends to pull the router into the workpiece.

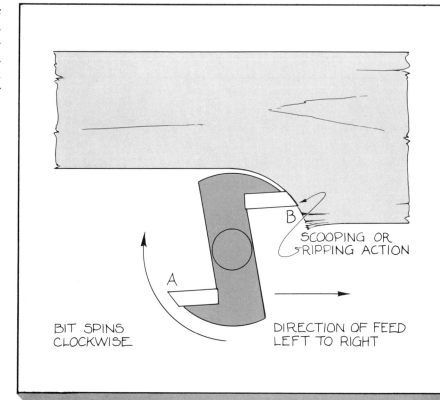

B

SCOOPING OR RIPPING ACTION

A

BIT SPINS CLOCKWISE

DIRECTION OF FEED LEFT TO RIGHT

which way a router is moved along a fence when routing within the body of a piece of wood? You can begin to understand the answer to this question by conducting a little experiment. With a ¾-inch (19-mm) straight cutting bit chucked in your router, set the depth of cut to ⅛ inch (3 mm). Without trying to control the router too tightly, begin to rout freehand from left to right (**fig. 3-25**). What happens? The router will naturally pull itself away from you in a gentle arc. Now rout from right to left. The router will move in an arc toward you in a mirror image of the first cut. This phenomenon is a function of the force and resistance you saw before in edge routing. **Figure 3-23** shows that the resistance met by cutting edge B forces the router from its left-to-right course.

The point is simple. As you rout along a

fence within the body of a piece of wood, the router should be moved from left to right. In this way, the router holds itself close to the fence. I am sure that many an inexperienced woodworker has missed this rule, only to find that the router wandered away from the fence. This principle applies to curved guides, both concave and convex, as well as to straight fences. **Figure 3-26** shows the proper direction for moving the router against curved guides.

Now how does all of this translate into practical terms? In edge routing, the standard maxim has been that one should always rout from left to right. It is true that the scooping and ripping action produced in left-to-right routing is easier and faster than the crosscutting action that results from right-to-left routing. In addition, there is the ad-

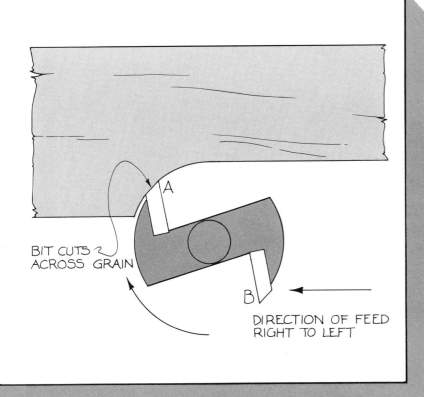

3-24: In right-to-left routing, cutting edge A encounters resistance, in what is essentially a crosscutting action. The resistance met at point A will then tend to push the router away from the workpiece.

BIT CUTS ACROSS GRAIN

A

B

DIRECTION OF FEED RIGHT TO LEFT

3-25: When freehand routing, the router will move in a predictable pattern. As the freehand router is moved left to right (A), the router pulls itself to the left, or away from the operator. When the router is moved right to left (B), the router pulls itself to the left but toward the operator.

3-26: The proper direction for moving a router against a curved (both concave and convex) guide.

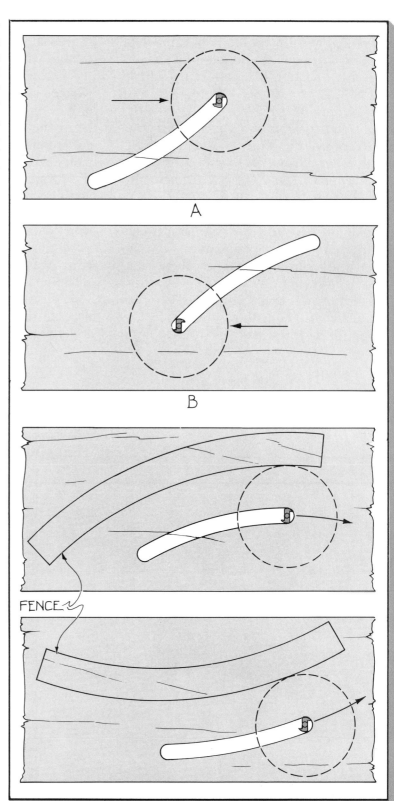

3-27: This cut was made with the router moving from left to right, but the oblique patterns of the grain caused the wood to chip.

vantage that the router is pulled into the workpiece. So why would one ever want to rout from right to left? One reason is that wood grain does not always run in a straight and uniform way. It might run straight, dart obliquely, and then swirl in irregular patterns. **Figure 3-27** shows how left-to-right routing can result in the grain tearing out. So, although the crosscutting action of right-to-left routing forces the router to work harder, it does result in a smooth cut without the risk of the grain tearing out.

In edge routing, therefore, the general movement of the router is from left to right, although one should be prepared to rout in the opposite direction when irregular grain would cause the grain to tear out. Not only should you examine the direction of the grain before you cut, but you should also pay attention to the sound of the router. Sensitivity to the sound of the router (and all woodworking tools, for that matter) cannot be stressed enough and is something that you will naturally develop in time. Your machine will maintain a high-pitched hum while cut-

ting smoothly, but the moment you hit the grain the wrong way, the sound will become more shrill—almost a scream. This noise is often accompanied by a snapping sound, which is a definite indication that the wood is tearing out. When this happens, stop the router. Examine the wood. Chances are the grain is swirling into the cut. The point I am getting at is that, as you approach the trouble area, you should pull the router away from the wood (pull toward you, but do not lift the router), and reenter beyond the swirling grain. Now rout from right to left. Hold on firmly, as the router may kick slightly. After cutting this section, you can then proceed from left to right, until another trouble area appears. Many cuts, however, can be made completely without any interruption of the left-to-right direction.

Here is another example of right-to-left routing. I often make finger grips for overlay doors with a ½-inch (12-mm) radius cove bit. The scooping action of left-to-right routing often leaves an unclean end to the cut, as shown in **figure 3-28,** when it should ideally end in a smooth taper. A way around this is to rout three-quarters of the finger grip from left to right and the last quarter (the edge) from right to left.

Now imagine setting a piece of ¼-inch (6-mm) glass into a frame (**fig. 3-29**). The notch or recess made to hold the glass is called a rabbet and is made with a ⅜-inch (9-mm) rabbeting bit with a ball-bearing pilot. I sometimes set the bit to a depth of ⅛ inch (3 mm) and make the first pass by routing entirely from right to left (counterclockwise around the inside perimeter of the frame). In this way, I need not worry about any grain tearing out on the top of the frame. After this pass, I set the bit to the full depth of cut and rout in a left-to-right direction around the inside of the frame.

3-28: The scooping action of left-to-right routing caused the grain to tear at the end of this finger grip. This tearing, as shown, could easily have been avoided if the router had been brought from right to left.

3-29: The generally prescribed routing pattern for rabbeting the interior of a frame is left to right, or clockwise, around the inner perimeter of the frame. This direction, however, can be reversed if swirling grain is causing the wood to tear out. Right-to-left routing will produce a cleaner cut, but care must be taken not to rout too deeply on any one pass.

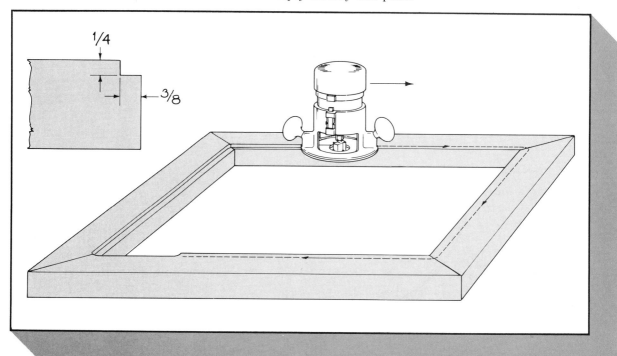

GENERAL ROUTING PROCEDURE

Routing requires a purposeful approach. How you stand, how deep you cut on a given pass, and how you start and stop a cut are all critical factors to successful routing. Let us examine these factors individually.

Setting Depth of Cut

I wish there were some simple, quantitative rules for setting the depth of a cut, but there are, unfortunately, too many variables involved—the size of the bit, the horsepower of the router, the species and amount of wood being cut, the sharpness of the bit, and so on. Pay attention to the sound and feel of your router when cutting (does it seem to be straining?); observe the quality of your cuts (is the wood burning?); and, in time, you will develop an intuitive sense for how deep you can cut. Generally speaking, be conservative. Routing a board in two passes hardly takes more time than routing in one pass. Your bits will thank you (they will stay sharper longer), and the quality of your cuts will be improved. Remember that putting a strain on your router by cutting too deep (or too fast) only serves to reduce the r.p.m., thereby reducing the number of cutting strokes per minute. Fewer strokes mean increased po-

tential for chipping. The Roman ogee profile shown in **figure 2-26** was made in two passes, but a carbide cutter in a 2¾-h.p. router was used. A larger ogee might have required three passes.

Making shallow cuts is particularly important when routing from right to left. Taking too deep a cut will almost inevitably result in the router kicking back, sometimes with great force.

When using the router table, you should make straight grooves as shallow as even ⅜ inch (9 mm) in two passes. The first pass is more difficult, since there is no place for the chips to escape.

Securing the Workpiece

Most router cuts can be made with the workpiece locked between two bench dogs (**fig. 3-30**). Bench dogs are blocks of wood that fit into square holes along the front end of the workbench; they are standard equipment on most workbenches. Always place a protective piece of wood between the bench dog and the workpiece. If you do not have such a bench, or the workpiece is too large for it, you will have to affix your wood in one of the ways shown in **figures 3-31** and **3-32**.

When edge routing, make sure that the workpiece hangs over the bench and, in general, see that no clamps, blocks, or securing strips of wood will be in the router's path. Odd-shaped pieces will often require custom blocking (**fig. 3-33**).

Beginning the Cut

There are several important things to remember before turning the router on. The cord of the machine should be placed around the back of your neck. This prevents a dangling cord from getting in your way during a cut. Place the router on the wood, but make sure that the bit is at least 1 inch away. This will prevent the router from kicking back when you turn it on, which can be dangerous to both you and your wood. Do not be over-eager. On most routers the on-off switch (usually a toggle switch) can be reached by the thumb while still holding on to the handles. On D-handle routers, the trigger switch is built right into the handle. Start the router

3-30: A small piece of wood protects the workpiece from the "bite" of a bench dog.

3-31: Affixing the workpiece to the bench with clamps.

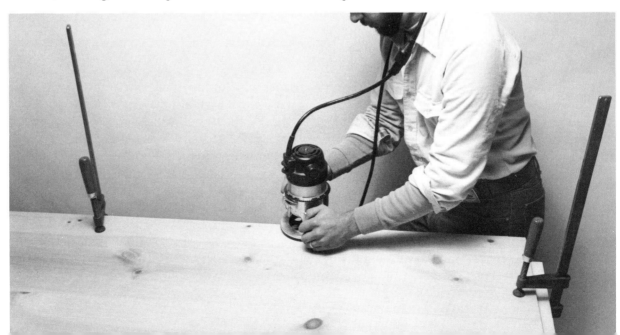

 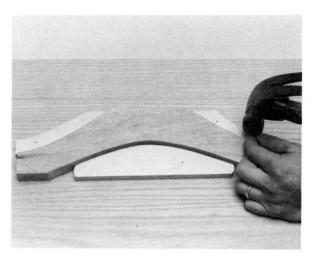

3-32 (left): If you have many pieces of the same size to rout, you can create a stabilizing frame by nailing scrap wood to the workbench. **3-33 (right):** Odd-shaped pieces require special blocking.

(it will jump slightly) and give it a few seconds to reach its full speed. Slowly enter the cutter into the wood.

I suggest pushing the router away from you as you proceed. In this way you can see any obstacle that is ahead of you (such as a clamp) that could cause an errant cut. Do not bear down against the workpiece too hard if you are controlling the cut with a piloted bit. Even a ball-bearing pilot can score the wood.

Rate of Feed

The rate of feed is the speed at which a router is moved. Knowing how fast to cut depends on the same variables as knowing how deep to cut. As such, this knowledge is also acquired with experience, and again I suggest that you be conservative. One router manual that I have seen states that routing too slowly is "as bad" as routing too fast. This is nonsense. Rout slowly. You will soon be able to

feel if you can go faster. If you rout around the perimeter of a board, for example, the sound of the machine will change as you rout crossgrain. If you are moving too fast, you will actually hear and feel the router straining and you will naturally slow down. Feeding too fast is a bad habit. It will often result in mistakes on the workpiece. And it will cause dull bits, sometimes a smoky shop, and usually a tired router.

You will want to go particularly slowly when edge routing from right to left, since kickback is a problem when routing in this direction. Hold the router a little more tightly than usual and be prepared for the router to kick (or jump) occasionally.

At the conclusion of any cut, shut the router off by reaching up with the thumb for the on-off switch. If the bit is free of the wood, lift the router up and away from the body and wait for the router to come to a complete rest (**fig. 3-35**). If the bit is still in contact with the wood, as in the case of a stopped dado, shut off the router, but do *not* lift it until the cutter has completely stopped. In either case, make the end of a cut a conscious part of the routing process; many an accident has occurred after a cut has been made.

PREVENTING ACCIDENTS

There is probably no aspect of woodworking about which I have stronger feelings than safety. An accident on the router or any other machine could have a dramatic effect on the rest of your life—and usually not one for the better. In one instant of carelessness, the joy of woodworking can be shattered forever.

In my years as a teacher, I have seen the full spectrum of personalities as they emerge in the use of tools and machines. Some people approach tools with timidity, others with aggression. Among even the most careful of beginners, however, I have noted a lack of fluidity, a slight awkwardness. It takes time for safe, proper use of the router to become second nature.

In the beginning, a very conscious effort to adopt the right attitude must be made. One has to self-impose a certain discipline. In checklist fashion you have to ask yourself: "Is the cutter locked in tightly?"; "Do I have the wood firmly clamped?"; "Am I taking too deep a cut?"; and so on. There should never be any question as to the safety of a particular cut. If any doubt lingers, do not make it.

Here are some of the more important safety guides to remember. Please read them with care.

● Unplug the router when changing cutters. Before you ever change a bit, the proverbial red light should flash; the machine should be disconnected. It is quite easy to hit the on button accidentally while wrestling with a stubborn cutter. Similarly, always make sure your router is turned off before you plug it in. If you install a switch on your router table, remember that the router switch itself is left in the *on* position. The problem occurs when you remove the router from the table and forget to switch it back off. You then plug the router in only to find that it is "live."

● Wear safety goggles (**fig. 3-34**). Safety

3-34: Safety goggles and ear protectors are a must. Flying chips of wood can easily scratch the cornea. Ear protectors dull the scream of the router and make routing more enjoyable. Notice that the cord of the machine has been placed over the back of the neck.

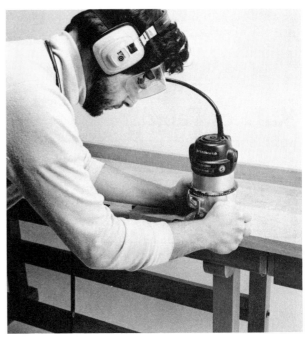

goggles are probably more important when using a router than when using any other tool, largely because of the large amount of dust sprayed during routing. Sawdust can be a real irritant to the eyes, while a flying sliver of wood can easily scratch the cornea. Perhaps more important is the fact that you want to have clear vision when routing; there are few things worse than being "blinded" during the middle of a cut. (Note: For photographic reasons, the author is not wearing eye goggles in some photographs.)

- The router cutter should always be at least 1 inch away from the wood to be cut before starting the machine. In order to cut properly, a router must first achieve maximum speed.

- Make sure the wood is firmly secured to the workbench. Double-check to see that clamps are tight. Along this line, all thumbscrews, clamps, and adjustments on the router itself should be secure.

- Always cut slowly. Cutting too quickly is suggestive of impatience, a dangerous frame of mind to have when routing. By working slowly, the chance of an accident occurring decreases markedly.

- Wear ear protectors. I can personally attest to the importance of this rule. For the first seven or eight years of my woodworking career, I wore no ear protectors. I also found that the constant exposure to loud machines was beginning to have a detrimental effect on my hearing.

- Pay attention to the sound of the router. Quality ear protectors will still allow you to do this. Throughout a cut, the router is giving you messages, mostly through sound; problems, for instance, are often revealed through a change in sound. As your sensitivity increases, you will find it easier to decode these messages. A change in sound could indicate a slipping bit. If the router sounds as if it is laboring (this sound will become familiar), it may

be telling you that your bit is cutting too deeply or perhaps that you are pushing the router too fast. A dull cutter will often tear the wood, resulting in a snapping sound. Dull cutters, by the way, will usually smoke. Smoking (don't worry, your router is not on fire) can also mean that you are taking too deep a pass or that the wood is very hard.

- Do not hesitate to shut off a router during a cut. If you suspect that something is wrong, do not complete the cut. Shut off the machine. Wait until the cutter stops completely before lifting the router, particularly on a cut such as a dado, where the slightest twist of the router could result in a ruined piece of wood. After the router has been stopped, analyze the cause of the problem. If you want to feel the cutter's edge for sharpness, unplug the machine.

- Be particularly careful about where you place your hands when using the router in the inverted position. In the inverted position (with or without the router table) your fingers can come quite close to the almost invisible bit (a cutter spinning at 20,000 r.p.m. is not easy to see). Your fingers should not be straining. They should be at once firm and relaxed. The moment they grip the wood too tightly, they have the potential to slip, and they should certainly be as far from the cutter as possible while still maintaining full control and pressure.

- Watch your movements at the end of a cut. When your cut is completed, reach up with your thumb and shut the machine off. You can either lift the router away from you and wait until the machine comes to a complete rest, as shown in **figure 3-35,** or you can hold the router on the wood and wait for it to stop. Never drop the router to your side after a cut while the bit is still spinning.

- Read the owner's manual that comes with your tool. This will provide specific

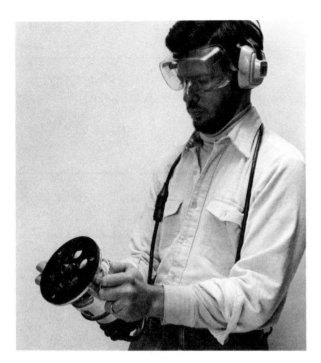

3-35: Many accidents occur after a cut has been completed. At the conclusion of a cut, either lift the router up and away from you and wait for it to come to a complete stop, or turn the router off while it is still in the workpiece, wait for it to stop, and then lift it out.

safety suggestions and information about the safety features of your particular router.

- Wear the proper clothing. Wear sturdy shoes with good, gripping soles. Shirts should be tight fitting; sleeves should never dangle. Ties, neck chains, jewelry, and so on should all be removed and long hair tied back.

- Make sure your tool is properly grounded. Your router cord will have a three-pronged plug. Your extension cord should have a three-pronged receptacle and the outlet should be properly grounded. Ideally, the circuit should be protected with a ground-fault circuit interrupter (GFCI) that guards against a circuit's overloading and overheating, while also protecting you against line-to-ground shock.

You could memorize all the safety rules ever written, but the real key to safe woodworking is attitudinal: it is the awareness of possible danger brought to bear each time a power tool is used.

4 The Router Table

Moving the router along a piece of wood can at times be awkward, particularly if your wood is small or narrow. By inverting the router and attaching it to the underside of a table, you have changed it from a portable tool to a stationary one. In essence, it has become a shaper. Wood that was previously too small to support a moving router is now easily pushed against the router table fence and onto the protruding bit. Repetitive cuts, such as for making tenons, are accomplished more efficiently, without the problem of clamping and unclamping the wood after each cut. With an adjustable fence and stop block, you can position wood quickly for cuts of pinpoint accuracy.

TABLE STYLES

Although router tables are commercially available, most woodworkers prefer to make their own. Commercial tables, as shown in **figure 4-1,** tend to be a little small. The fences are usually flimsy, and it is difficult to clamp around the bent edges of the metal top. What are the alternatives? I have seen woodworkers content themselves with simply attaching their routers to a flat piece of wood. On the other end of the spectrum are the more elaborate homemade, floor model router ta-

bles. In between, I have seen just about every idiosyncratic variation, including one adapted by a friend from a child's school desk. All he had to do was mount the router and add a fence.

Another solution is to attach your router to the underside of your table saw (**fig. 4-2**). This necessitates bringing the top of the table saw to a machine shop and having a flat machined on the underside. This flat must be perfectly parallel to the top; if it is not, the

4-1: This is a router table sold by Stanley.

4-2: The router can be mounted to the underside of a table saw. This type of router table takes advantage of the saw's fence and miter gauge.

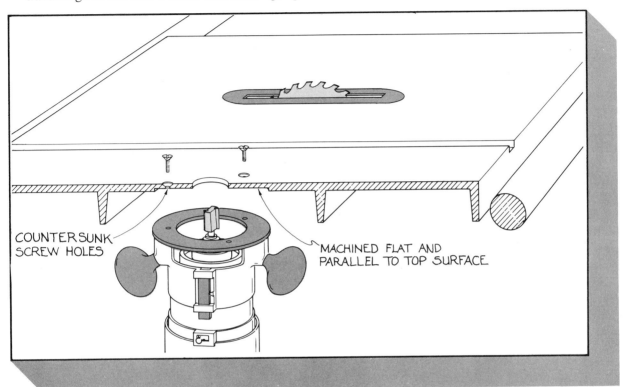

COUNTERSUNK
SCREW HOLES

MACHINED FLAT AND
PARALLEL TO TOP SURFACE

4-3: Another alternative is to mount the router on the extension table of the table saw. This obviates the need to bring the table-saw top to a machine shop.

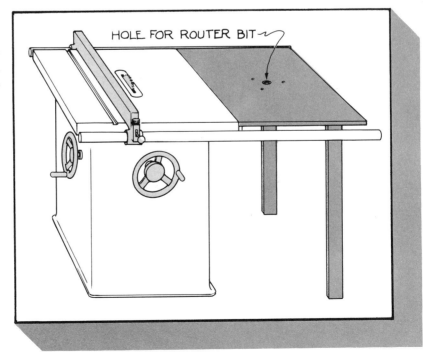

HOLE FOR ROUTER BIT

bit will not protrude square to the table. The chief advantage of this system is that you can utilize both the saw's adjustable fence and its miter gauge. In addition, you have not taken up any more space in your workplace. The chief disadvantage of the table-saw router table is that, with the router in use, the table saw cannot be used. In a commercial shop, this would be a serious problem. If you like the basic concept of a table-saw router table, but do not want to go to the expense and trouble of machining the top, mount the router on a table-saw extension table (**fig. 4-3**).

If space permits, I advocate the use of a free-standing router table. I allude to my own router table, as shown in **figure 4-4**, throughout this chapter, and plans for its construction can be found in **figure 4-5**. Simple or elaborate, however, a router table is not a complicated affair.

4-4: My own floor model router table.

The Table Top

Above all else, the top must be perfectly flat. Three-quarter-inch plywood covered with a white plastic laminate is probably your best bet. Pencil marks show up easily against the white surface and can be easily wiped off. Use a matte finish laminate, as it has a better grip than a gloss finish, and be careful to lay the contact cement on in two even but thin coats. Any lumps in the cement will telegraph to the working surface. Other acceptable materials for the top include any smooth, ¾-inch (19-mm) plywood, phenolic (the same material used to make the sub base), and ¼-inch (6-mm) tempered masonite glued to a piece of particleboard. For more information on applying the laminate to the table, see Chapter 9, *Laminate Trimming*.

Although you select a top for its flatness, it is really the frame of the router table that serves to hold it flat. If the table you build does not have four aprons, add battens, such as these shown in **figure 4-5.**

In terms of size of the top, there is no fixed rule. It can be as small as the commercially built ones, approximately 13 inches × 18 inches (33 × 45.7 cm) or as large as your workplace will allow. The top of the floor model table in **figure 4-4** is 23 inches × 34½ inches (58.4 × 87.6 cm).

Positioning and Mounting the Router

The router should be positioned according to the size of the table and your own personal comfort. On a small router table, as shown

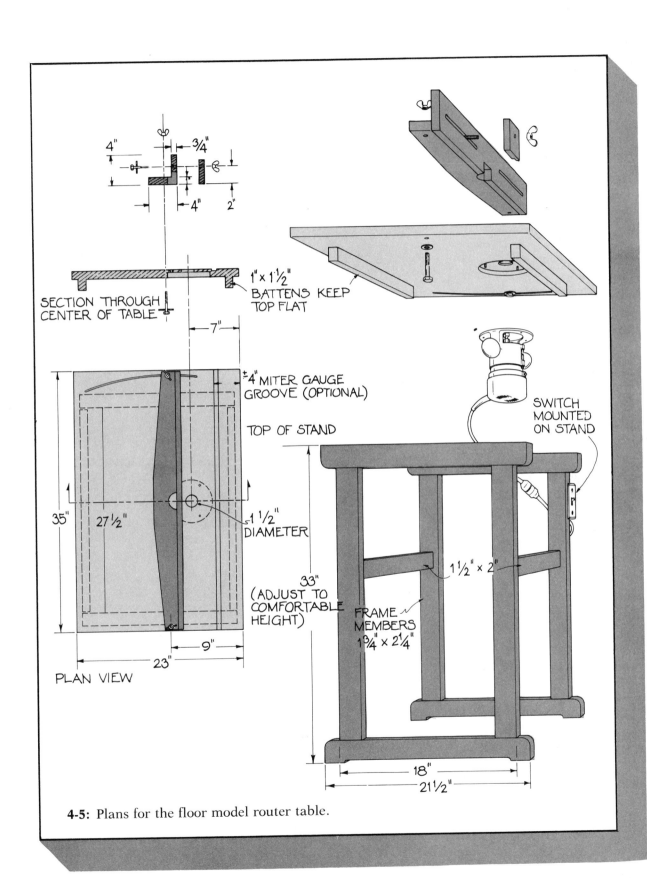

4"
3/4"
4"
4"
2"

SECTION THROUGH
CENTER OF TABLE

1" x 1 1/2"
BATTENS KEEP
TOP FLAT

7"

±4" MITER GAUGE
GROOVE (OPTIONAL)

TOP OF STAND

35"

27 1/2"

1 1/2"
DIAMETER

33"
(ADJUST TO
COMFORTABLE
HEIGHT)

SWITCH
MOUNTED
ON STAND

1 1/2" x 2"

FRAME
MEMBERS
1 3/4" x 2 1/4"

9"

23"

PLAN VIEW

18"

21 1/2"

4-5: Plans for the floor model router table.

4-6: I bring this small router table with me onto the job site. This table has a miter gauge, which my large router table does not. The fence is just a straightedge held by two C clamps.

in **figure 4-6,** the router is mounted on the "far" side, thereby providing maximum table surface to support the wood to be cut. This is not a good place to mount the router on a large table, however, since you would be leaning over constantly to make your adjustments. **Figure 4-5** shows a router mounted more to the "near" or operator's side. Most cuts can be made comfortably this way. If you need more space to accommodate a large board, turn the fence around and work from the other side of the table.

While methods for mounting the router base to the underside of the router table can vary as a function of different router models, the most universal method is to attach it with the same machine screws used to hold the sub base on. Begin by removing the sub base. The sub base then becomes a template for marking the position of the screw holes (**fig. 4-7**). Drill the holes large enough so that the screws slip through easily. Countersink so that the screw heads sit just below the surface. You will soon see that the screws are

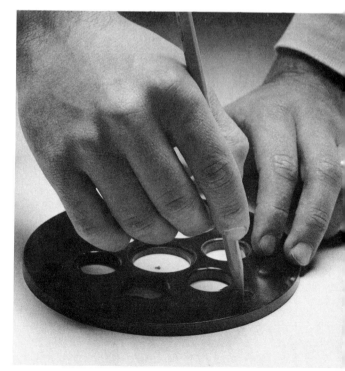

4-7: To mount a router, begin by marking the position of the sub-base screw holes on the top of the router table.

4-8: Mounting a router to the router table.

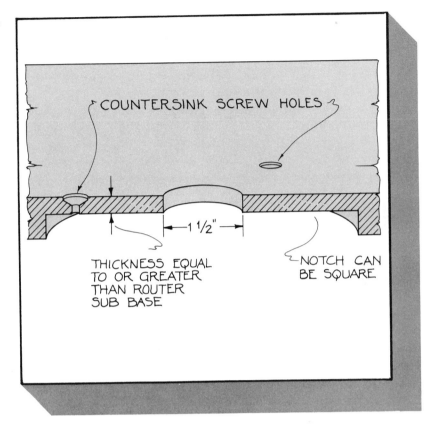

COUNTERSINK SCREW HOLES

← 1 1/2" →

THICKNESS EQUAL TO OR GREATER THAN ROUTER SUB BASE

NOTCH CAN BE SQUARE

not long enough to go through the thickness of the top and into the router. You could use longer screws, but you would still be saddled with the problem that the thickness of the top has reduced the router bit's depth adjustment by at least ½ inch (12 mm). I prefer to rout a recess in the underside of the top so that the thickness of the top at that point is equal to or slightly larger than the thickness of the sub base (**fig. 4-8**). The recess need not be round nor fit tightly around the router. You will also have to drill a 1½-inch (38-mm) hole to allow the bits to protrude.

Although the router base is easy to remove and reattach, I suggest that you purchase an additional router base and leave it permanently affixed to the table. Another convenience and safety option is an on-off switch for the router table (**fig. 4-9**). This surface-mounted box has an outlet (into which the router cord is plugged) that is controlled by

4-9: A surface-mounted switch saves you the trouble of having to feel around blindly for the motor switch every time you want to start or stop the machine.

a switch. This eliminates feeling around blindly for the switch every time you want to start or stop the machine. A word of caution: using a switch will mean that the router switch is left in the *on* position. Remember, then, to switch it off when taking the router off the router table.

Fences

Ideally, the router table should be outfitted with several different fences, each designed for a different purpose. In its simplest form, a fence is a straight block of wood that is clamped to the table top (**fig. 4-6**). The next logical step is to use an adjustable fence (**fig. 4-10**). Notice that the fence is L-shaped and made from two pieces. The separate pieces decrease the likelihood of the fence warping and provide a space for attaching the nuts

and bolts. The fence must have a notch cut into it at the point of the cutter, since there will be many times when you will want to expose only a portion of the bit. The notch also serves as an outlet for the chips and dust made during the cut.

Two other excellent (but expensive) alternatives are to purchase fences designed for other machines. A table-saw fence, for example, makes an excellent router-table fence. Even better is a shaper fence, which not only has fine-tuning (vernier) adjustment, but is also designed to permit the infeed and outfeed portions of the fence to move independently of each other.

It is important to note that a fence need *not* be parallel to the edge of the table or "square to the bit" in order to produce a straight cut (**fig. 4-11**). A fence can be clamped at any angle. The only critical factor is the distance between the outside edge of the bit and the fence itself. If, in making a groove, for example, the distance between the bit and the fence is ¼ inch (6 mm), the groove will be ¼ inch (6 mm) away and par-

4-10: An adjustable router-table fence. Plastic handles can be substituted for the wing nut shown.

4-11: A fence need not be parallel to the edge of the table ("square to the bit") to make a straight cut. The exception to this rule occurs when a miter gauge is being used in conjunction with a fence. In that case, both the fence and miter-gauge groove must be parallel.

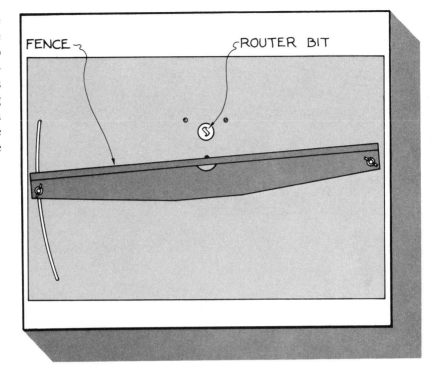

4-12: A high auxiliary fence is indispensable for many router table operations.

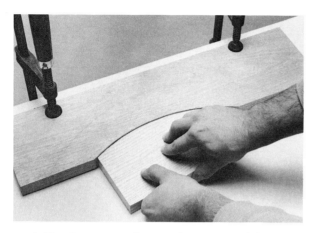

4-13: One type of curved segmented fence.

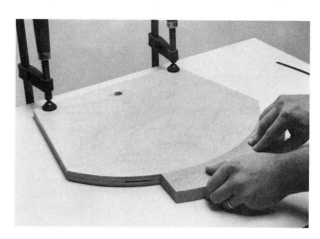

4-14: Another curved segmented fence.

allel to the edge of the board. Only when you use a miter gauge in conjunction with a fence does the fence have to be "straight " (parallel to the miter-gauge groove).

An auxiliary "high" fence (as shown in **fig. 4-12**) is particularly helpful in providing support when routing pieces of wood on end. It should be at least 6 inches (15.2 cm) high. Curved segmented fences used for curved work are shown in **figures 4-13** and **4-14**.

Fences can be made with built-in stop blocks, to use when starting and stopping cuts. One system is shown in **figure 4-15**. Notice that the stop block is rabbeted at the

end. This provides a place for sawdust to collect so that it will not get caught between the block and the workpiece. By the same token, the stop block is raised off the table to allow sawdust to pass underneath.

A miter gauge, as shown in **figures 4-1** and **4-6**, is another good router-table accessory. Just borrow one from another machine and rout a groove for it. Other excellent safety accessories include spring catches (also called "spring hold downs") and safety guards. The ones made by Rockwell adapt beautifully to homemade router tables. See *Sources of Supply*.

4-15: The stop block on this fence is easily adjusted by loosening and tightening a wing nut on the other side of the fence. The end of the block is rabbeted to make a place for sawdust to collect.

5 Edge Routing

The majority of routing that you will do will probably be along the edges of boards. This may include the routing of decorative edges and moldings, which may need several successive passes with different bits or settings to achieve the desired effects. Edge routing may also include rabbeting, which is a groove or notch cut along the edge of a board, as may be needed for the insertion of wood or glass. Or the router can be used as a jointer, smoothening edges that are to be joined. The router can also be used for splining, in which case grooves are routed into the edges of two boards before insertion and gluing of a thin spline of wood into grooves.

DECORATIVE EDGE ROUTING AND MOLDING ROUTING

While other router bits provide the means of construction and joinery in your work, the decorative edge bits add the artistry. Whether it is in the beading on a door frame or the classic elegance of a Roman ogee profile, these bits will bring subtlety and grace to your woodworking.

Exercise your imagination in the use of decorative edge bits. The Roman ogee edge, for example, can be used in several different ways to produce very different effects (**figs. 5-1** and **5-2**). There may not be a router bit available to you that can produce a full half-round edge, but a quarter-round (or rounding-over) bit used on each side of a board will produce the same effect (**fig. 5-3**). In addition, by exposing varying amounts of one bit, different shapes can be achieved (**fig. 5-4**). Remember that pilots can be interchanged to modify the shape of a bit. A smaller pilot, for example, can change a rounding-over bit into a beading bit (**fig. 2-14**).

Bits are sometimes used in combination with one another, as is the case with a rule

5-1 and **5-2:** On the top of a cabinet, the Roman ogee edge can be placed up or down to produce very different effects.

5-3: This edge was produced by using a quarter-round (rounding-over) bit on each side.

joint (**fig. 5-5**). This "joint" is made with a complementary (they have the same radius) set of bits—a rounding-over bit for the table and a cove bit for the dropleaf.

Decorative edge bits have pilots, but using a standard edge guide in conjunction with the pilot will give the router greater stability and keep the pilot from marking the wood. If you are routing one edge of a board only, you want to be sure that the router does not "turn the corner" at the beginning and end of the cut. Before starting the cut, engage the front end of the fence (edge guide) against the workpiece and concentrate on keeping pressure at this point (**fig. 5-6**). At the end of the cut, do just the opposite and keep the pressure at the back of the fence. If the one edge that you are routing is a crossgrain edge, the cut should be made in two stages to prevent the grain from tearing out (**fig. 5-7**). If you are routing all four, follow the sequence shown in **figure 5-8**.

Routing around the perimeter of a round

5-4 (left): One bit, in this case a Roman ogee, can produce quite different shapes by cutting at different depths. **5-5 (right):** A rule joint on a reproduction Shaker harvest table. This joint is easily routed with a cove bit and rounding-over bit that have the same radius.

piece presents a particular problem in terms of grain direction. Although the prescribed routing pattern is left to right, or in this case, counterclockwise around the circle, **figure 5-9** shows that it may be necessary to rout from right to left (clockwise) along the edges

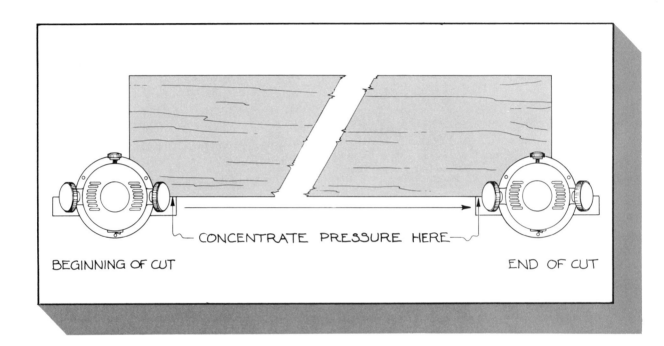

CONCENTRATE PRESSURE HERE

BEGINNING OF CUT   END OF CUT

5-6 (above): Keeping pressure on the right points of the fence will prevent the router from "turning the corners" at the beginning and end of a cut.

5-7: Routing across the grain of a board in two stages will prevent the grain from tearing out.

STOP

FOR LAST INCH OR SO ROUT FROM RIGHT TO LEFT

5-8: This is the safest sequence for routing the perimeter of a board in order to prevent any kind of cross-grain chipping.

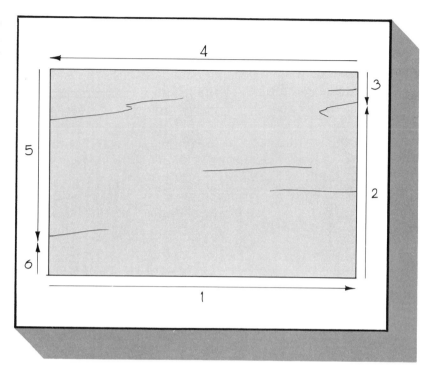

5-9: Routing the edges of a circle can result in grain direction problems. It may be necessary to rout from right to left in quadrants A and D to prevent the grain from chipping.

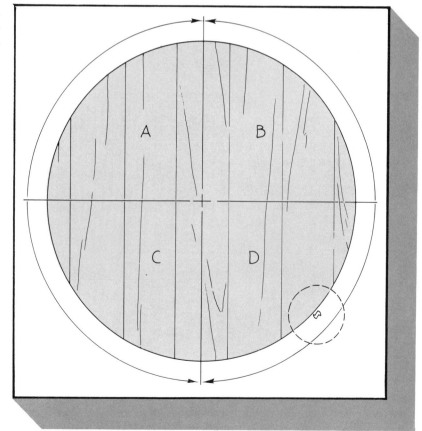

of quadrants A and D, to prevent the grain from tearing out.

Moldings are best made on the router table. Do not cut the moldings to dimension before routing the shape. First, cut your wood to a width of no less than 1½ inches (38 mm). Rout the profile on each side of the piece (**fig. 5-10**). After all the routing is completed, cut the moldings to final size on the table saw. More elaborate and curved moldings are typically made on an overhead router, as shown in Chapter 12.

Raised Panels

Raised panels, as shown in **figure 5-11,** can be formed with bits designed expressly for this purpose. Large panel-raising bits require a heavy-duty router (at least 2 h.p.) and should be used with a router table. Do not try to make the profile in one pass.

5-10: Rout moldings on a piece of wood no less than 1½-inch (38-mm) wide. The moldings will be ripped to final dimension afterward. This method is much safer than trying to rout narrow pieces. Note, as shown, that profile is routed on each side of the piece.

5-11: A panel-raising bit.

RABBETING FOR WOOD AND GLASS

A rabbet is a groove or notch along the edge of a board. Among other things, rabbets are used to hold cabinet backs. The router is the logical choice for making back rabbets because the rabbet must often begin and end at specific points as shown in **figure 5-12,** something that is virtually impossible to do on the table saw. **Figure 5-13** shows the routed dimension for setting in a typical ¼-inch (6-mm) back. Although one might assume that a rabbeting bit would be the choice for this cut, this is not the case. A rabbeting bit has too fixed a dimension; backs are rarely an exact ¼ inch (6 mm), and the rabbet has to be adjusted accordingly.

I do the preliminary setup and test cut with the router inverted and locked in a vise (**fig. 5-14**). Remember, your router must have

5-12: Rabbets for cabinet backs usually stop short of the ends of the workpiece.

the necessary flanges in order to do this. I suggest using a ½-inch or ¾-inch (12- or 19-mm) straight cutting carbide bit. Set the depth of cut to at least half the thickness of the wood. Next, position the edge guide so that ¼ inch (6 mm) of the bit is exposed. Make a test cut, but instead of checking the accuracy of the cut with a ruler, actually set a small piece of the back to be used into the rabbet (**fig. 5-15**). It should set in about ¹/₃₂

5-13: The dimensions for setting in a typical ¼-inch (6-mm) back.

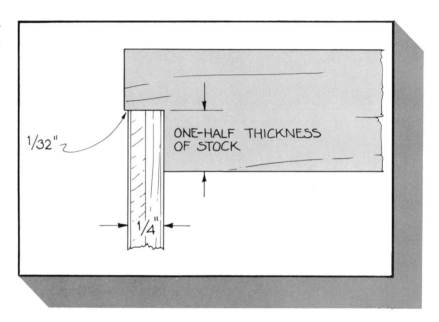

5-14: As shown, the preliminary test cut is set up on an inverted router.

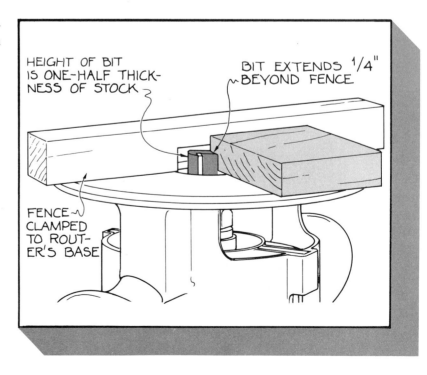

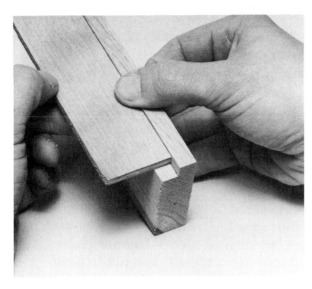

5-15: Do not check the accuracy of your test cut by measurement. Instead, fit a scrap of the actual back into the rabbet.

inch (.79 mm). Remove the router from the vise. Only the test cut is done with the router inverted; the actual routing of your pieces is done with a moving router.

If a rabbeting bit is not used for back rabbets, it *is* used for rabbeting a frame that is to receive glass or a mirror (**fig. 3-29**). This type of rabbeting is typically done after the frame (door, table top) that holds it has been glued. The frame should be planed carefully so that all surfaces are uniform and flat. A ⅜-inch (9-mm) rabbeting bit provides a good support shoulder for the glass. Depth of cut will, of course, be determined by the thickness of the glass being used. The direction of feed for this cut is detailed in **figure 3-29.** The rabbet bit will, in this case, leave rounded corners, which will later have to be chiseled square (**fig. 5-16**).

5-16: A rabbeting bit leaves rounded corners, which you will have to chisel square.

EDGE JOINTING

Solid wood boards are rarely found with smooth, straight sides and the woodworker is left with the task of trueing or jointing these edges. This can be done easily with the router in a number of different ways.

Short boards—less than 36 inches (91.4 cm), for example—are readily jointed on the router table, which you should set up to function almost exactly like a jointer (**fig. 5-17**). You can use any size straight cutting bit so long as its cutting length exceeds the thickness of the boards you are working with. I use a ½-inch (12-mm) carbide bit with a 2-inch (50-mm) cutting edge. A ½-inch (12-

mm) shank is preferable because of the considerable lateral stress exerted on the bit.

The edges of most boards are shaped either concave or convex. It is much easier to joint a concave edge because it is better supported by the fence. If the curve is severe, there will be a substantial amount of material to remove, but it is never advisable to rout more than ⅟₁₆ inch (1.5 mm) at a time. Crosscut your boards to the shortest possible lengths to minimize the curve and use the table saw (I do it freehand) or saber saw to remove any large amounts of wood (**fig. 5-18**).

As with a jointer bed, the infeed and out-

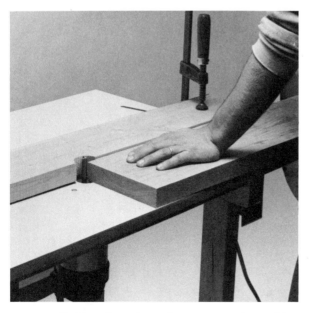
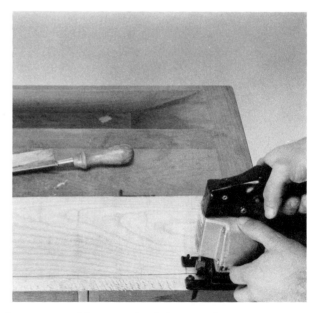

5-17 (left): Short boards are jointed quickly on the router table. **5-18 (right):** Significant amounts of wood should be sawed off prior to routing.

feed sections cannot be on the same plane or the wood would be unsupported on the outfeed section as the edge is routed (**fig. 5-19**). You can outfit your router table with a fence in which the infeed and outfeed sec-

tions move independently of each other, or you can simply construct a jointing fence, in which the infeed and outfeed sections are permanently offset (**fig. 5-20**). To make a jointing fence, start with a straightedge. Cut

5-19: The infeed and outfeed sections of a jointer fence cannot be on the same plane, or the wood would be unsupported on the outfeed section as the edge is routed.

5-20: An offset jointing fence. The outfeed section of the fence is flush with the bit so that the wood is firmly supported while it is jointed.

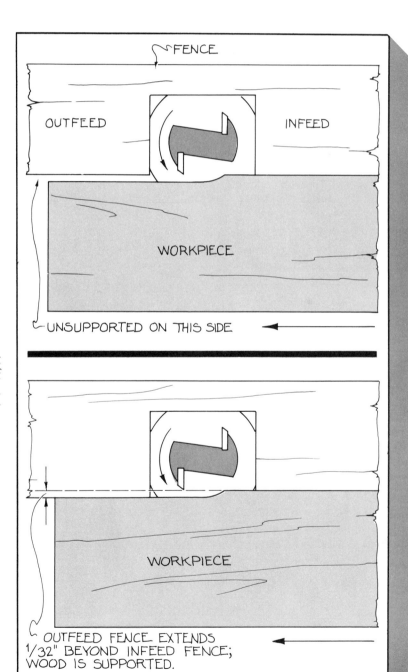

a notch in the fence to delineate the infeed and outfeed sections. This notch will surround the bit. With a table saw or router and a trim guide, remove 1/32 inch (.79 mm) from the edge of the infeed section. Clamp this fence to your router table and, as with a jointer bed, align the outside cutting edge of the bit flush with the outfeed section. Run a piece of wood through and notice that an exact 1/32 inch (.79 mm) is removed, and the board is also firmly supported by the outfeed fence (**fig. 5-20**). Make several passes until the edge is perfectly straight. I have two jointing fences—one offset 1/32 inch (.79 mm) and the other 1/16 inch (1.5 mm).

Long boards are more easy to joint with a straightedge and moving router. Perhaps the easiest method is to use your dependable friend, the trim guide. After sawing away any large amounts of wood, simply clamp the guide to the work and rout the edge straight (**fig. 5-21**). If you have a nonconcentric router, make sure that you keep the router in the same position throughout each cut.

To avoid problems with a nonconcentric router, use a flush-trimming bit in conjunc-

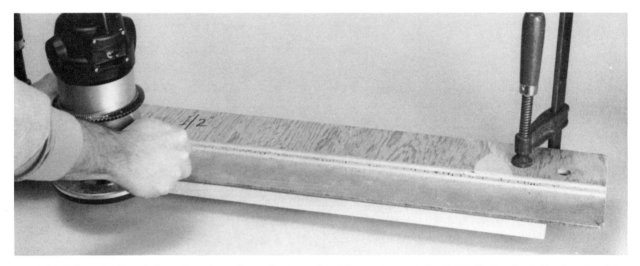

5-21 (above): Straightening the edge of a board with a trim guide. **5-22 (below):** A flush-trim bit riding along a straightedge can also be used to joint a board.

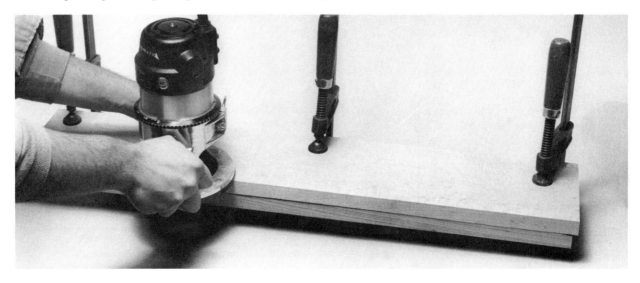

tion with a straightedge (**fig. 5-22**). Clamp the straightedge to the underside of the workpiece, setting it at a distance equal to the amount of wood you want to remove. The pilot rides along the straightedge and transfers its straightness to the workpiece.

When I want to joint a board and also to join it edge to edge with another board, I use a virtually foolproof technique to put a perfect "seam" between two boards every time. You will have to build the board jointing table as shown in **figures 5-23** and **5-24.** The

5-23 (above): Here is the jointing table for edge-to-edge joining. **5-24 (below):** Use these dimensions for building the jointing table shown above.

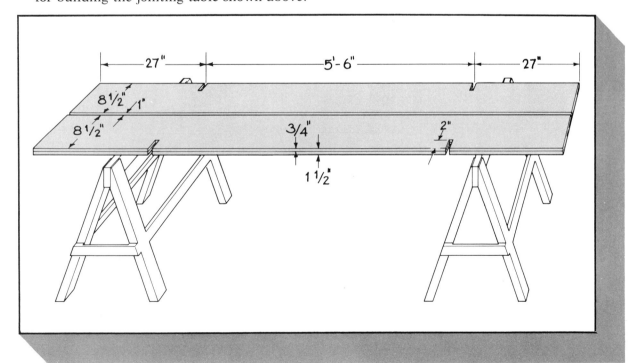

5-25: In routing piece A, the router is moved along the straightedge so that between $\frac{1}{32}$ inch (.79 mm) and $\frac{1}{16}$ inch (1.5 mm) is removed.

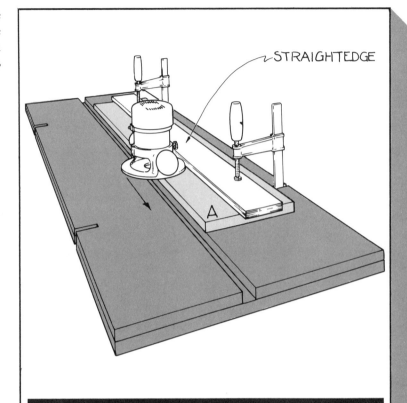

5-26: Without removing piece A or touching the setup in any way, piece B is routed in a second pass.

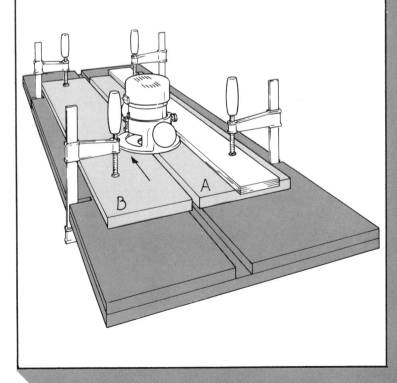

table should be between 6 and 8 feet (5.5 and 7.3 m) in length. You will need a straightedge 3 inches (7.6 cm) wide and a little longer than the table. I made my jointing table from particleboard so that its own weight prevents it from shifting during a cut. The trough in this fixture allows the bit to cut below the boards, and it is also an outlet for the considerable amount of sawdust made while routing.

Place the table on two sawhorses. Let's say you want to joint a perfectly mated edge between boards A and B. Position piece A on the jointing table so that the edge to be jointed hangs over the trough (**fig. 5-25**). Position the straightedge and clamp it in position so that the bit will rout from between 1/32 inch and 1/16 inch (.79 and 1.5 mm) off the board. Parallel jaw clamps extend in deeply and are probably the best clamps for holding the wood flat, although I sometimes use deep-throated bar clamps instead. Before you cut, examine your bit for nicks. Even a small nick will leave a fine ridge on the routed edge and prevent the boards from coming together. Make this cut as many times as necessary until the bit makes contact along the full length of the board. This edge is now true.

Without removing the clamps or touching piece A in any way, position piece B so that the distance between the two boards is 1/16 inch (1.5 mm) less than the width of the cutter (**fig. 5-26**). Clamp piece B in place, taking care to position the clamps so that they will not interfere with the router. Guiding the router against the same fence, bring it back through, in the opposite direction of the first pass. Only piece B is being cut on this pass. The theory is very important: you are routing two boards "simultaneously," but in two passes and against one fence. Any deviation

5-27: Putting clean edges on these tiger maple boards was almost impossible to do on the jointer. They were, however, easily mated on the router-jointing table.

5-28: This zebrawood box has an undulating band (not an inlay) of walnut that runs around all four sides of the box.

from straightness in the fence will transfer to both pieces. In other words, it does not matter if the fence is not perfectly straight, because the two boards have been cut against the same edge and will automatically be mated.

If the router comes away from the fence when cutting piece A, the edge of piece A will be cleaned up on the second pass. If, however, the router drifts on the second pass, you are in trouble. Swirling grain can sometimes pull the router away, so do not be afraid to bear down and against the fence. Although

the breeze from the router will blow away most sawdust as you rout, sawdust can sometimes get between the router base and the fence. This, too, can ruin a cut.

You can take this basic routing concept a step further by using curved or irregularly shaped fences. These edges can create dramatic effects, particularly if contrasting woods are used (**figs. 5-27** and **5-28**). Begin by cutting the curved fence and transferring its shape onto the two pieces to be mated (**fig. 5-29**). Bandsaw their shapes, cutting just to the outside of your pencil lines. The rules

5-29: Routing mating curves. The basic approach is essentially the same as in edge-to-edge jointing shown on preceding pages.

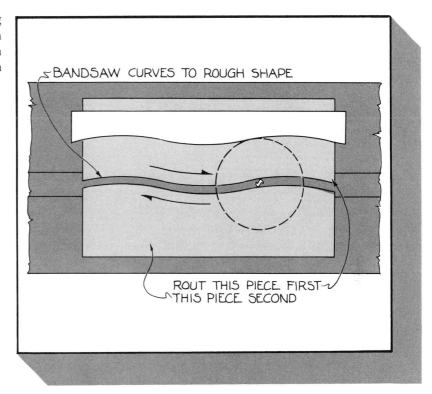

BANDSAW CURVES TO ROUGH SHAPE

ROUT THIS PIECE FIRST
THIS PIECE SECOND

for routing are exactly the same as for straightline routing, except that you should make the lightest passes possible to minimize chipping. In spite of some basic mathematical impossibilities built into this routing method (the shape of the workpiece, for example, will not be an exact transference of the shape of the fence), the system will work, so long as the curves are gentle. The serpentine curves will attract attention.

SPLINING

Boards glued edge to edge can be strengthened with dowels, splines, or a tongue-and-groove joint. Splining is the insertion of a thin strip of wood between two boards and is the fastest of the three methods above (**fig. 5-30**). Spline grooves are best made with a router (as opposed to a table saw) for two reasons. First, the splines are often blind—that is, they do not come through to the ends of the boards. With a router, you can start and stop the grooves at will (**fig. 5-31**). The

second reason is that wood is often twisted or curved, and the router can be set to follow the contours of the wood.

There are two basic methods for splining. The first uses the router table. For ¾-inch (19-mm) stock, use a spline ¼ inch (6 mm) thick by ¾ inch (19 mm) wide. In each board, the groove will be about ¹⁄₃₂ inch (.79 mm) deeper than ⅜ inch (9 mm). With an average router (1½ h.p.), this depth should be made in two passes. In addition, the grooves should

5-30: Strengthening a joint with splines.

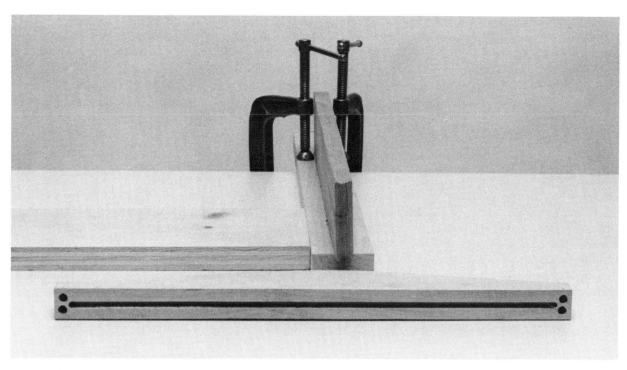

5-31 (above): Routed spline grooves can be started and stopped at will. **5-32 (below)**: Splining on the router table. A featherboard helps to keep the workpiece against the narrow guide.

5-33: Splining with a slotting cutter.

be parallel to the surfaces of the wood, so that the boards being splined together can work to keep one another flat. A curved board will not run flat against a standard fence, so, in order to keep the groove parallel to the surfaces, you will have to use a narrow guide, such as the one shown in **figures 5-31** and **5-32.** Always keep the ''good'' face against the guide so that even if this groove is not centered, it will always be the same distance from the face on all boards. Using a featherboard will give you more control **(fig. 5-32).**

The second method, using a slotting cutter, is easier and more accurate, but requires a heavy-duty router (at least 2 h.p.). Chuck the cutter, set the depth, and spline away **(fig. 5-33).** I find this type of splining invaluable in many different situations. When, for instance, I have to cut and joint many countertops on the job site, I bring precut splines with me and simply rout the grooves where I need them.

One problem with slotting cutters is that they cut quite deep, often deeper than necessary. Even a 2- or 3-h.p. router has to work hard. One solution is to insert a larger pilot,

5-34: Most slotting cutters cut deeper than necessary and can put a strain on even high-powered routers. A good solution is to substitute one of these large nylon bearings (available from the Fred M. Velepec Co.) for the standard ball bearing.

which will cause the slotting cutter to cut more shallowly. With a shallower cut, it also becomes possible for a smaller router (1½ h.p., for one) to perform slotting operations. Nylon bearings, which can be used instead of pilots, are particularly good for this job, since they are available in large sizes and are less expensive than steel bearings **(fig. 5-34).** They are available from the Fred M. Velepec Company. (See *Sources of Supply.)*

6 Secrets of Joinery

At the heart of cabinetmaking lies the art of joinery, which, as the name implies, is the craft of fitting wood together. With the addition of glue (it is glue, not nails and screws, that holds furniture together), a properly fitted joint becomes almost impossible to break. So, in making joints, you are trying to create as many glue surfaces as possible. It is in the cutting of joints that the router's tremendous versatility comes to the fore.

I am often asked when and why joints should be made with the router and not the table or radial arm saws. This is an important question. A professional cabinetmaker will, generally speaking, use a power saw (most often the table saw) to fullest advantage when making joints. The table saw permits fast setups, and its powerful motor can breeze a blade through any board. When making a rabbet on the table saw, for example, only two small cuts are required before a substantial piece of wood falls away. The router, on the other hand, has to "chew" its way through all that wood. A table saw can make hundreds of rabbets without dulling the blade, whereas a router bit would tire much more quickly. In addition, using the router often means that you must clamp and unclamp the wood over and over again, which can be very time consuming.

On the other hand, the router possesses certain advantages over the power saws. For one thing, it produces a cleaner, squarer cut. Clean, square cuts are particularly important when the joinery will be exposed. The router can make stopped grooves; the table saw, because of the large radius of the blade, leaves elongated curves. In addition, there are certain joints, or parts of joints, that are difficult to cut on a power saw. A mortise, as shown in **figure 6-31,** is a clear example of this type of joint. Dovetails, when not cut by hand, also fall within the province of the router. Very large work can be too cumbersome to move along a table saw; a moving router can be passed above the oversized piece. Finally, some woodworkers simply do not have table or radial arm saws and must rely exclusively on their routers to make joints.

When making joints, it is imperative that each cut be preceded with a practice or test cut on a piece of scrap wood. It is not enough to measure for the depth of cut or just position the fence. A test cut will tell you exactly how accurate that measurement was. The test piece should be the exact same thickness as the wood being worked on; ideally, they would be cut from the same board. If you were to make a tenon, as shown in **figure 6-46,** and the test piece was thicker than the actual wood, you would cut all the tenons only to find that they all fit too loosely. Making a test cut can also reveal any problems that might arise during a cut. It is better to make a mistake here than on a beautiful piece of wood.

THE RABBETED JOINT

The rabbet-and-groove joint is a strong corner joint used extensively in carcase construction. The basic configuration of the joint is shown in **figure 6-1.** One rule of thumb is that the thickness of the tongue on the rabbeted piece (B) is typically one-third the thickness of the stock, which on the ¾-inch (19-mm) stock shown would be ¼ inch (6 mm). When routing this joint on ½-inch (12-mm) stock, however, I usually bend this rule and still cut a ¼-inch-thick (6-mm) tongue.

Quarter-inch tongues require ¼-inch (6-mm) grooves, and in production routing a ¼-inch (6-mm) bit holds up better than a smaller bit. The length of tongue (B) and consequently the depth of the groove in piece A is equal to half the thickness of the material with which you are working.

When using the router table, make the groove first. With a ¼-inch (6-mm) straight cutting bit inserted in the router, set the depth of cut. If your router is 1½ h.p. or

6-1: The dimensions for a rabbeted joint in ¾-inch (19-mm) stock.

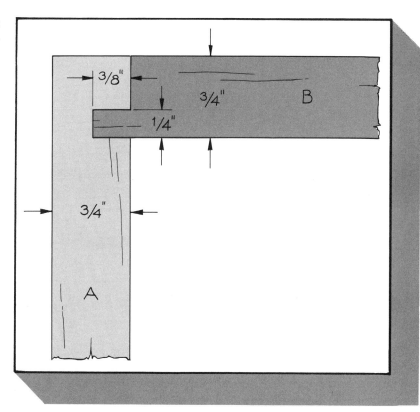

3/8"
3/4" B
1/4"
3/4"
A

6-2: Distance "d" is equal to the thickness of piece B.

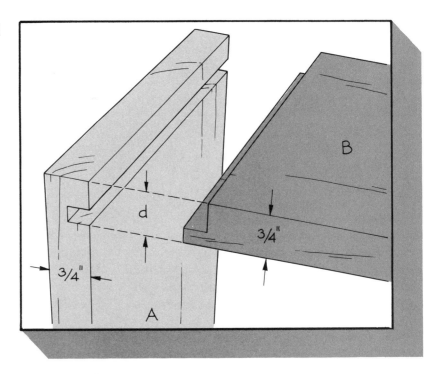

6-3: Using piece B (or a test piece of equal thickness) to position the router table fence. Notice that the bit protrudes fractionally beyond thickness of piece B.

6-4: When the rabbeted joint is completed, the top of piece A sits fractionally above piece B. The tongue is also shorter than the depth of the groove, to ensure a tight outer seam.

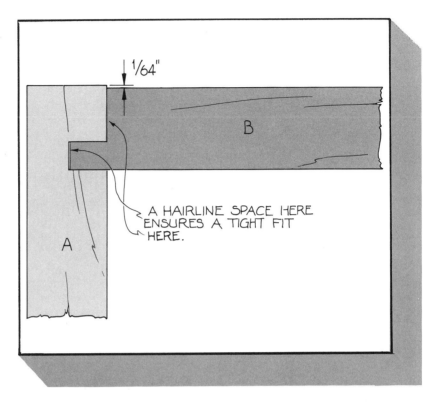

greater, the cut can probably be made to the depth of ⅜ inch (9 mm) in one pass.

Figure 6-2 shows that distance "d" is equal to the thickness of piece B. To set the router table fence, then, make the distance from the fence to the outer edge of the router bit equal to the thickness of piece B, which is ¾ inch (19 mm) (**fig. 6-3**). Rather than measure this distance as ¾ inch (19 mm), however, hold a test piece (the *exact* same thickness as piece A) against the fence and to the outer edge of the bit. This is more direct and accurate. Notice also that the bit is protruding approximately ¼₄ inch (.39 mm) beyond the edge of the test piece. This ensures that after the joint is completed, as shown in **figure 6-4,** the top of piece A will protrude just above the top of piece B. It is certainly easier to flush the top of piece B with a flush-trimming bit than it would be to plane the full top of piece A (**fig. 6-5**).

With the depth of cut set to ⅜ inch (9 mm), or ³⁄₁₆ inch (4.7 mm) if you are making the groove in two passes, and the fence posi-

6-5: Using a flush-trimming bit.

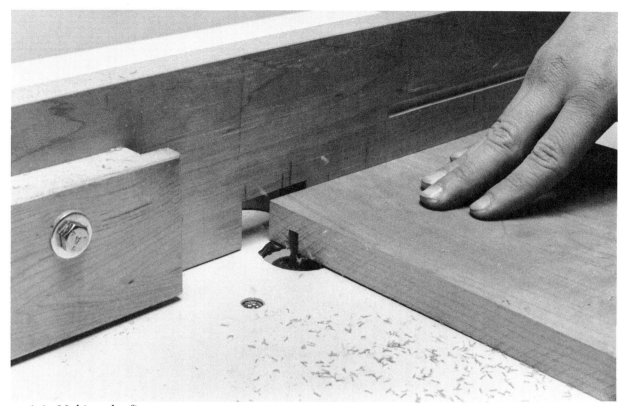

6-6: Making the first test cut.

tioned as shown in **figure 6-3,** make your first test cut (**fig. 6-6**). To determine whether the groove has been accurately placed, hold the test piece against it (**fig. 6-7**).

To cut the rabbet, change to a ½-inch or ¾-inch (12- or 19-mm) straight cutting bit. The exact size is not critical, so long as the bit exceeds ⅜ inch (9 mm), which is the length of the tongue. By adjusting the router table fence in or out, you can expose as much or as little of the bit as you choose, thereby creating rabbets of different sizes. The height of the bit should be set so that the tongue that remains is ¼-inch (6-mm) thick. The position of the fence is such that the length of the tongue is equal to the depth of the groove (**fig. 6-8**). In point of fact, the tongue should be cut fractionally less than the depth of the groove to allow for glue and to ensure that the joint is tight all along its outside, visible portions (**fig. 6-4**). In addition, firm,

downward pressure is mandatory when making both the grooves and rabbets, to guarantee that the bit is cutting to its full depth. If the wood is very curved or wavy, as shown in **figure 6-9,** it may be necessary to make the cuts by using a moving router (**fig. 6-10**) rather than an inverted router. As long as the grooves and rabbets are cut to the proper dimensions, however, the curvature in the wood will ultimately be straightened when the cabinet is glued. I also recommend moving the router along the wood when cutting the rabbeted joint on very large pieces, which are sometimes awkward to cut on the router table.

Finally, there may be times when you want to make the rabbet-and-groove joint blind. You can achieve this easily by stopping the groove approximately ¼ inch (6 mm) short of the forward edge of the wood and cutting the tongue back with a handsaw (**fig. 6-11**).

6-7: Testing to see if the groove was properly positioned in relation to the end of the board.

6-8: The tongue of the joint is defined by cutting a rabbet. The thickness of the tongue is equal to the width of the groove, and the length of the tongue is equal to the depth of the groove.

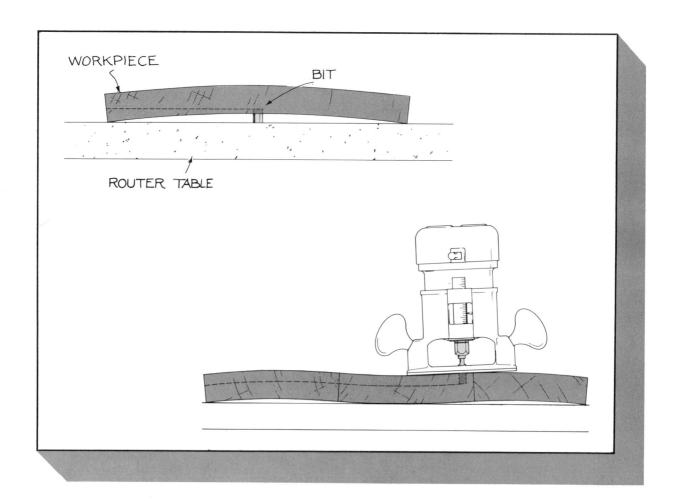

WORKPIECE

BIT

ROUTER TABLE

6-9 (top): A warped or curved board will not rest flat on the router table, resulting in a tongue of unequal thickness or a groove of unequal depth. **6-10 (above right):** A moving router will follow the contours of a curved board, resulting in cuts of consistent dimension. **6-11 (right):** A blind rabbeted joint.

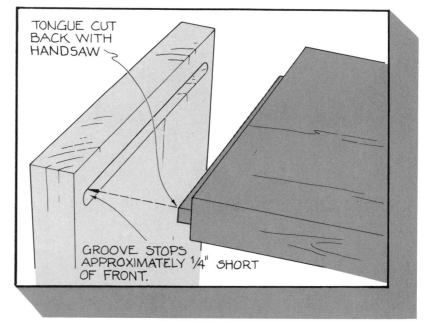

TONGUE CUT BACK WITH HANDSAW

GROOVE STOPS APPROXIMATELY 1/4" SHORT OF FRONT.

A variation of the basic rabbeted joint, and one that is excellent for joining drawer sides and faces, is the locking corner joint, as shown in **figure 6-12.** To the uninitiated, this joint might appear a little perplexing, but it is really quite simple to make. The first cut to be made is groove A in the drawer face (**fig. 6-13**). The groove is ¼-inch (6-mm) wide, so chuck a ¼-inch (6-mm) straight cutting carbide bit (**fig. 6-14**). The depth of the cut is equal to the thickness of the drawer side, which, for our purposes here, you can assume is ½ inch (12 mm). The groove, then, will be ½-inch (12-mm) deep, although you should do it in two passes. Set the depth of cut, therefore, to ¼ inch (6 mm).

The distance from the face of the drawer to the groove is plus or minus ³⁄₁₆ inch (4.7 mm). The exact size is not critical and can be adjusted for aesthetic reasons. Set the distance between the router table fence and the bit at ³⁄₁₆ inch (4.7 mm). (Do not forget your test cuts.) With the face of the drawer held against the router table fence, make the first pass. Now, raise the cutter so that its height is equal to the exact thickness of the

6-12: This locking corner joint is a strong and attractive choice for drawer construction.

6-13: Here are dimensions for groove A for the drawer face in the sequence for making the locking corner joint, continued next page.

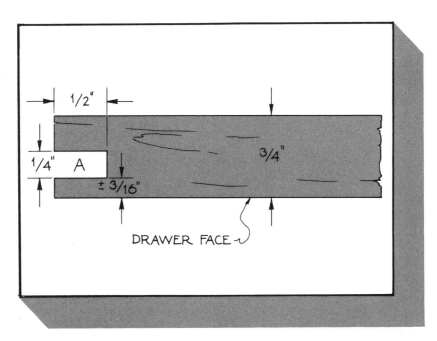

drawer sides, ½ inch (12 mm) in this case.

The second cut removes plus or minus ¼ inch (6 mm) off tongue B, as shown in **figure 6-15.** The drawer face is now completed. All that remains is to cut a single groove in the drawer sides (**fig. 6-16**). The distance between the fence and the bit—the same ¼-inch (6-mm) bit—is ¼ inch (6 mm), which is the width of groove A. Make the groove very slightly deeper than the length of the tongue B, to allow room for glue and to ensure that the joint sets properly. Tongue B, it should be noted, will be thicker than ¼ inch (6 mm), so that groove C will have to be widened until it fits over this tongue.

6-14,15,16: These drawings continue the sequence begun on the previous page for the locking corner joint.

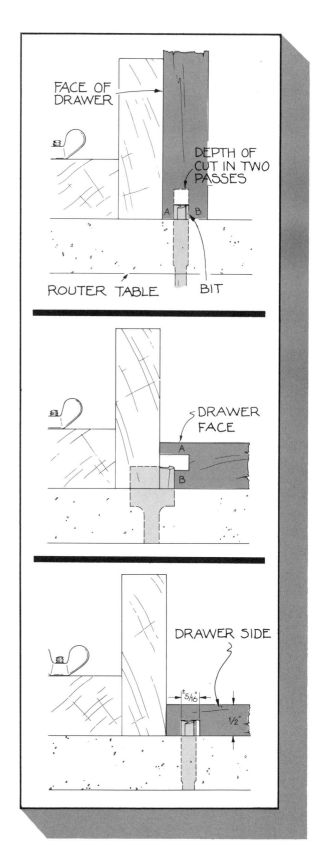

THE TONGUE-AND-GROOVE JOINT

The tongue-and-groove joint is quite similar to the rabbet joint, except that the tongue is defined by two rabbets instead of one (**fig. 6-17**). It is somewhat stronger than the rabbet joint because of its two shoulders, but is not used often as a corner joint. Rather, the tongue-and-groove joint is used more in making horizontal shelves and vertical partitions. The depth of the groove is equal to half the thickness of the wood. The thickness of the tongue is typically one-third the thickness of the stock.

There are several suggested methods for making this joint. Begin by making the groove. Assuming ¾-inch (19-mm) stock, insert a ¼-inch (6-mm) straight cutting bit into the router, and set the depth of cut to ⅜ inch (9 mm). The grooving jig shown in **figure 6-18** will assure accurate grooving with minimal measuring. Distance "d" is equal to the width of the router base. The length of the jig is optional, but I recommend a length of between 36–48 inches (.9 and 1.2 m) because there will be times when you will rout two wide boards simultaneously. An additional 16-inch (40.6-cm) jig is convenient for smaller work. The two long rails should be at least 2½ inches (63.5 mm) wide, so as not to deflect when routing. Piece A squares the jig to the edge of the board, and the groove in its center is a quick alignment guide. Once you have marked the grooves in the work, it is simply a question of positioning the jig and then doing the routing.

6:17: The tongue-and-groove joint is used in case construction to join horizontal shelves and vertical partitions

Sometimes I prefer to cut the grooves for interior shelves and partitions *after* a cabinet has been glued. Consider the recessed partitions in **figure 6-19**. Note that the sides and the top of the cabinet are mitered. The bottom (B) is joined to the sides with a stopped rabbet-and-groove joint, therefore the top and bottom are of different lengths. To make grooves for the partitions in this case would require careful measuring and repetitive marking out. After all, a slight misalignment of the grooving jig would certainly result in an out-of-square partition.

6-18: The grooving jig assures accurate grooving with minimal measuring. Clamping workpieces together for the purpose of grooving them also ensures that the grooves will be aligned when the pieces are separated.

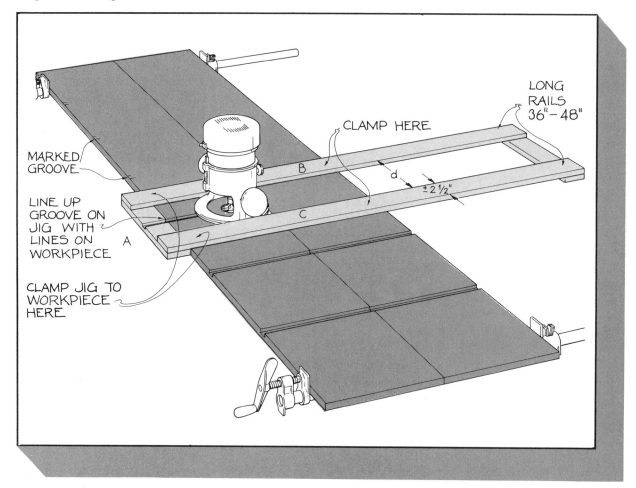

An easier approach would be to cut the grooves after the basic carcase is glued. Mark the position of one groove (**fig. 6-20**). Insert a ¼-inch (6-mm) bit into the router and measure the distance from the outer edge of the bit to the outer edge of the router base. Subtract this figure from the distance (*A* refers to the distance from the cabinet side to the first groove line). Cut a piece of wood to this size and clamp it to the cabinet bottom. This piece will be the guide for routing all four grooves. **Figure 6-21** shows the direction for moving the router. When you move the router from left to right, the router will pull itself in toward the fence; if the router were moving from right to left, it would have a tendency to wander away from the fence. In order to start the groove on the left side of the cabinet, you must first make a plunge cut, although you need not have a plunge router to do this. The groove will be set back approximately ¼ inch (6 mm) from the forward edge of the partition. Mark a line on the cabinet to indicate the beginning of the groove. You will be routing to this line by eye. Hold the router tight against the fence. Tip it up so that the bit is at least ½ inch (12 mm) off the workpiece in the back of the cabinet. **Figure 6-21** also shows the direction for moving the router when grooving to the

6-19: Because of differences in joinery, top A and bottom B are of different lengths. It will be easier to cut the grooves for the vertical partitions after the cabinet has been glued.

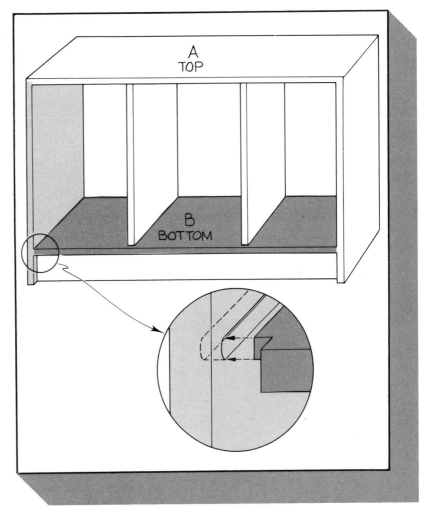

6-20 (below left): Marking a groove and positioning the router guide. When measuring for the height of the partition (B), measure in the corners and not in the center. **6-21 (below right):** Proper direction of the router. See also **Fig. 6-22.**

MARK OUT THE POSITION OF ONE GROOVE

RABBET ALLOWS GUIDE TO SEAT FIRMLY IN CORNER

right side of the cabinet. Turn the cabinet over and rout the other two grooves.

Note: Make sure there is no sawdust in the corners of the cabinet that would prevent the guide piece from seating firmly against the side. In a production routing situation, you would rabbet the end of the guide piece to avoid this problem.

To cut the partitions to the correct length, first measure distance as shown in **figure 6-20**. To this dimension, add the depths of the upper and lower grooves. This figure represents the length of the partition. Cut the tongues by rabbeting both sides of each end of the partition (**fig. 6-22**). Since the tongues in this example are blind, complete the joint by cutting the front of the tongue back with a handsaw.

When using this joint in plywood cabinets, it is usually quite easy to slide the partitions in. Solid wood, however, may be slightly curved, making it necessary to clamp battens to the wood in order to slide the partitions into position. If the joint is not blind and the tongue and groove is visible from the front, the fit can be eased by planing the tongue with a rabbet plane. Be careful not to plane the front, visible part of the tongue. You would then slide this partition into position from front to back.

The tongue-and-groove joint is also used for joining breadboard ends (**fig. 6-23**). The end cap, which holds the groove, should be chosen for its straightness. Using a slotting cutter or a ¼-inch (6-mm) straight cutting bit, make the groove ½ inch (12 mm) deep and ¼ inch (6 mm) wide.

The tongue is made by rabbeting both sides of the wood (as previously described), with the exception that the cuts are made against the grain (**fig. 6-24**). You will notice that routing across the grain causes a slight tearout (**fig. 6-25**). This can be prevented by scoring the joint with a marking gauge before cutting. If you do not want to take the time to do this, the hairlike projections caused by crossgrain routing can be pared later with a sharp chisel.

The tongue-and-groove joint for breadboard ends should not be so tight that it would prevent the boards from expanding and contracting freely. For aesthetic reasons, however, the joint must fit close at the ends. To guarantee a pleasing appearance, I rout the tongues for a close fit and ease the tightness along the joint with a rabbet plane.

6-22 (continued from previous pages): Cutting the tongues. The length of the partition is equal to B in **fig. 6-20** plus the depth of the upper and lower grooves.

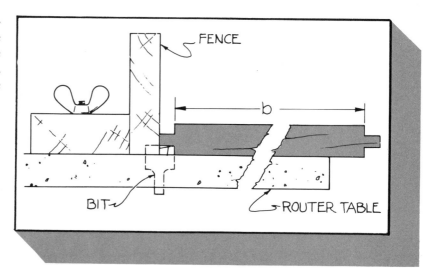

6-23: A breadboard end on a traditional table.

6-24 (below left): Cutting the tongue on the tabletop. The router fence regulates the length of the tongue. **6-25 (below right):** Routing across the grain can cause a slight tearing of fibers. This can be minimized by scribing with a marking gauge before routing. A sharp chisel, however, will quickly clean up the edge.

THE DADO JOINT

The dado joint, as shown in **figure 6-26,** is used most commonly for joining horizontal shelves and vertical partitions. Making a dado with a router is usually done by clamping a straightedge directly on the workpiece and moving a router along it. To ensure that the dadoes line up exactly on both sides of a cabinet, cabinetmakers sometimes clamp the two sides together, clamp a straightedge over both pieces, and rout the groove (**fig. 6-27**).

Although cutting dadoes against a straightedge is acceptable, two difficulties can arise. The first problem is that if the router comes away from the fence at any point during the cut, you have a ruined piece of wood. The router handle could accidentally hit a clamp during the cut, for example,

6-26: A dado joint.

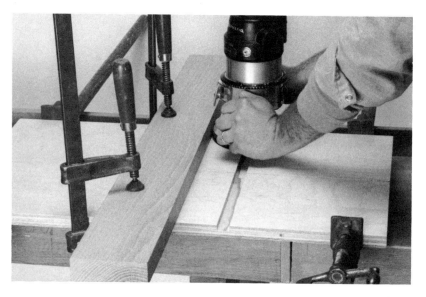

6-27: Pictured here are the two sides of a small bookcase being routed at the same time.

and cause the router to swerve. Second, wood is rarely the exact thickness it is supposed to be. Three-quarter-inch (19-mm) plywood is usually ¹⁄₃₂ inch (.79 mm) to ¹⁄₁₆ inch (1.5 mm) less than ¾ inch (19 mm). If you use a ¾-inch (19-mm) bit to cut a dado for this wood, the groove would be too sloppy. Theoretically, you could use a ⅝-inch (15.8-mm) bit, make one pass, move the fence, and make another pass. This approach is too messy, however, particularly if there are many dadoes to cut.

A simple dado jig, as shown in **figure 6-28,** solves both of these problems. The jig is almost identical to the grooving jig shown in **figure 6-18.** In the dado jig, however, distance "d" is larger than the router base. By regulating distance "d" with thin strips of veneer, you can cut dadoes up to ⅛ inch (3 mm) larger than the bit and any size in be-

tween. Imagine that you want to rout a ⁹⁄₁₆-inch-wide (14.2-mm) dado. Assuming that you do not have a ⁹⁄₁₆-inch (14.2-mm) bit (that would be too easy), you will have to use a ½-inch (12-mm) bit. Clamp the jig in position. As long as piece C fits snugly against the workpiece, you can be assured of a 90-degree cut. Clamp a ¹⁄₁₆-inch (1.5-mm) shim along fence B, rout fence A, and rout back along fence B. The dado should be exactly ⁹⁄₁₆ inch (14.2 mm), but you should make a test cut first anyway.

If the shelf fits too tightly into the dado, you can regulate its width even further by substituting a ¹⁄₃₂-inch (.79 mm) strip (I use veneer strips) for the ¹⁄₁₆-inch (1.5-mm) piece.

Another method for making dadoes is to use the trim guide. Simply position the jig along the line of cut, regulate the depth of the bit, and rout away (**fig. 6-29**). You will

6-28: A simple dado jig. Distance d is ⅛ inch (3 mm) greater than the width of the router base. This allows you to cut dadoes that are larger than the width of the bit.

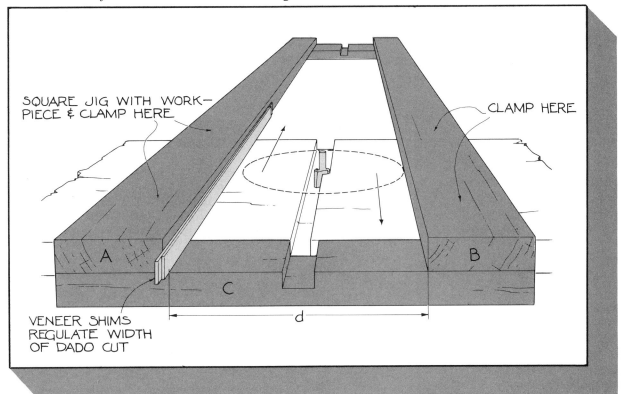

SQUARE JIG WITH WORK-PIECE & CLAMP HERE

CLAMP HERE

A

B

C

d

VENEER SHIMS REGULATE WIDTH OF DADO CUT

have to project the router bit an extra ¼ inch to compensate for the thickness of the guide itself. This is the fastest way to make dadoes, since there is no measuring, but it does not solve the problem of odd-sized dadoes.

On small pieces of wood, it is best to cut dadoes on the router table. Certainly on very small pieces, this is the only way (**fig. 6-30**). If you decide to use the miter gauge in conjunction with the miter table fence, it is imperative that the fence be parallel to the miter gauge groove.

6-29 (left): The trim guide is good for fast dado setups, since no measuring is involved. Simply clamp the trim guide along the line of cut, regulate the depth of cut, and rout. **6-30 (right):** Cutting dadoes on narrow wood is best done with a miter gauge, which is barely visible beneath my hand. Do not forget that the router-table fence must be parallel with the miter-gauge groove.

THE MORTISE-AND-TENON JOINT

For those who think that woodworking is a recent development, let them be reminded of the beautiful furniture produced by ancient Egyptian craftsmen, complete with turnings and inlaid veneers. The mortise and tenon was included in their repertoire of joints. The applications of this joint are innumerable, but it is used most extensively in door, table (leg to apron), and general frame construction. For our purposes, I will discuss three basic types of mortises and tenons: the closed mortise and tenon; the open mortise and tenon, both of which are shown in **figure 6-31**; and the through mortise and tenon (**fig. 6-42**).

Generally speaking, since it is easier to regulate the thickness of the tenon for a good fit, it is best to make the mortise first. Most mortises are cut on the router table. This obviates the need for clamping and unclamping the work for each cut. Position the fence so that the mortise is centered in the wood. Because mortises are cut fairly deep, they usually have to be cut in two or more passes. If I am cutting mortises that are deeper than, say, ¾ inch (19 mm), I usually drill out a large part of the material first on the drill press. This alleviates any strain on the router.

The only real question in setting up a mortise on the router table is how to start and stop the cut. In routing a closed mortise, the most controlled method is to use two stop blocks (**fig. 6-32**). The first block is clamped high on the fence. The workpiece is held tight against this block and is slowly lowered onto the bit. Notice that the back end of the workpiece is planted firmly on the router table and that the workpiece itself is pivoted down from this point. This ensures more control than if you were to simply drop the wood down. The workpiece is then advanced forward until it makes contact with the second stop block. If the pieces you are working with are wide, you will probably have to substitute a high auxiliary fence for your standard router-table fence.

Another way to start the cut is to clamp a

6-31: On top, an open mortise; on bottom, a closed mortise.

block at the rear of your workpiece (**fig. 6-33**). (This system presupposes that your pieces are relatively short.) If pinpoint accuracy is not critical on a particular cut, you can begin your cut by aligning the wood with a pencil line and advancing forward to the stop block.

When cutting mortises at either end of a workpiece and using these techniques, always keep the same side of the wood against the fence. In this way, if the bit is not perfectly centered on the wood, or if the wood varies slightly in thickness from piece to piece, the mortise will always be the same distance in from the face (**fig. 6-34**). In other words, you should not rout one mortise and simply flip the piece around to mortise the other end. After setting up the first mortise, as shown in **figure 6-32,** and running through one end of each of your pieces, you will have

6-32 (above): Starting and stopping a closed mortise. A starting block is clamped high on the auxiliary fence. The workpiece is lowered from this point and is fed until it hits the second stop block. **6-33 (below):** Starting the cut with a stop block at the rear of the workpiece. This method is effective when the workpiece is fairly short.

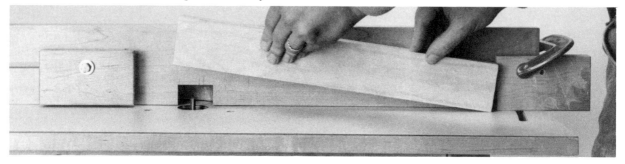

6-34 (below): By keeping the same side of the wood against the fence when cutting each mortise, distance "d" will always be the same.

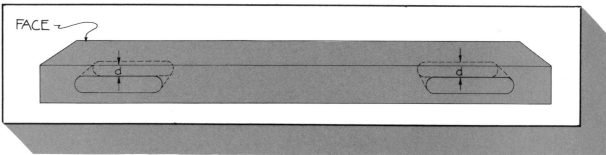

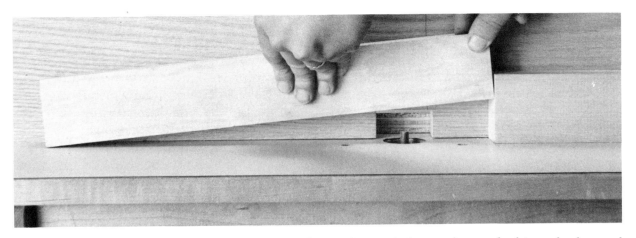

6-35: Cutting the back end of a workpiece. The workpiece is lowered onto the bit and advanced forward until the end comes in contact with a pencil line on the fence.

to make a secondary setup to mortise the other end (**fig. 6-35**).

Mortising a table leg, as shown in **figure 6-36,** also requires a two-part setup. **Figure 6-37** shows an open mortise being made (one that comes through to the end of the wood). The top of the leg is advanced forward until it comes in contact with the stop block. In **figure 6-38,** the top of the leg is held against the "near" stop block, lowered onto the bit, and advanced forward. In this way, both mortises are the same distance in from the outer edge of the leg.

6-36: The open mortises of a table leg.

6-37 and **6-38:** Cutting open mortises on a table leg. The two mortises have to be routed from two different positions to ensure that the mortises are the same distance in from the outer leg edge.

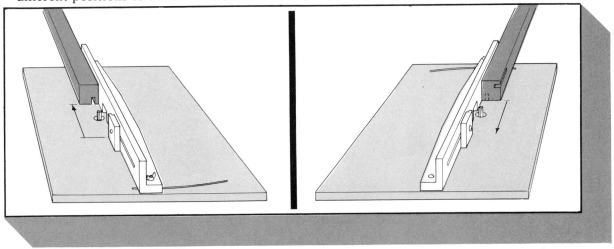

Plunge Mortising

Tage Frid, noted woodworker and author, has outlined another excellent system for mortising, using a plunge router. Frid's method utilizes the homemade mortising fixture shown in **figure 6-39.** The router (Frid uses the Makita 3600B) is set on top of the fixture and is guided along it by the auxiliary fence. The movement of the router and consequently the length of the mortise are controlled by two adjustable end stops.

Dimensions for the mortising fixture are given in **figure 6-40.** Frid cautions that dimension A should not be more than 3¼ inches (8.2 cm), or the router will not rest on the two side pieces. It is also imperative that the bottom of the fixture be perfectly square to the sides. If not, all of your mortises will be slightly angled.

You will have to use scraps of wood to bring your workpiece almost flush with the top of the side piece. The workpiece is then clamped to the side of the fixture. Place the router against the left-hand stop and hold the fence tight against the fixture. Release the lock lever and lower the bit into the wood. The depth of cut (on the Makita 3600B) is controlled by one of two depth adjustments. When the cutter reaches the preset depth, reengage the locking lever and move the router forward until it makes contact with the right-hand stop.

If I am making long and deep mortises, as, for example, on house doors, I use an even more direct system. After marking the mortise, I drill as much wood as I can. Because the stock for these doors is a minimum of 1⅛ inches (27 mm) thick, the router has enough bearing surface to rest directly on the wood.

6-39: Tage Frid's mortising fixture. The router fence rides along the side of the jig. The distance that the router travels, and consequently the length of the mortise, is controlled by the two end stops. See plans, **fig. 6-40.**

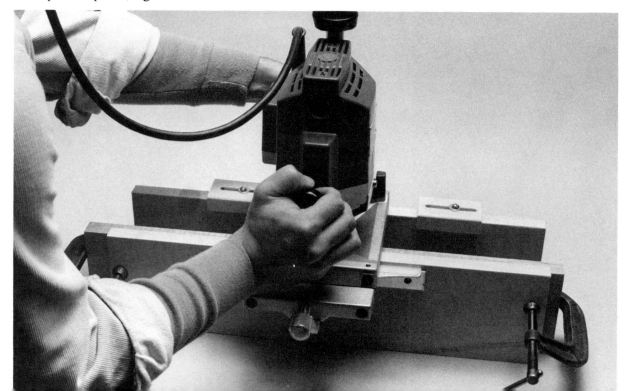

6-40: Plans for the Tage Frid mortising fixture.

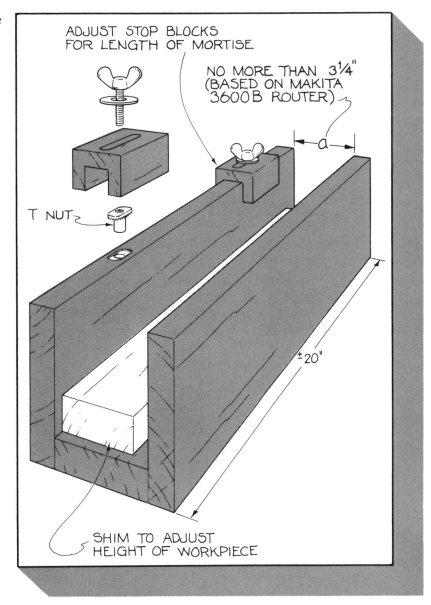

ADJUST STOP BLOCKS
FOR LENGTH OF MORTISE

NO MORE THAN 3¼"
(BASED ON MAKITA
3600B ROUTER)

a

T NUT

±20"

SHIM TO ADJUST
HEIGHT OF WORKPIECE

The auxiliary fence also helps to stabilize the router. Next, I position the bit just above the mortise, plunge, and then rout (**fig. 6-41**). I rout from one end of the mortise to the other by eye. With most of the mortise wasted by drilling, two passes will easily clean up the opening. This technique, by the way, does not require the use of a plunge router—any router will do just fine.

The Through Mortise

Another type of mortise is one in which the tenon comes all the way through the mortised stock (**fig. 6-42**). A simple wood pattern

6-41: Routing long, deep mortises with the router resting on the workpiece.

and flush-trim bit can reproduce these mortises over and over again. On a piece of ¼-inch (6-mm) plywood, mark the mortises as you want them. Drill and chisel them square. This is your pattern (**fig. 6-43**). Lay the pattern on your workpiece and mark the mortises. Drill the bulk of the wood. Next, you should clamp the pattern back onto the workpiece. A flush-trim bit, riding along the template, will then reproduce the mortises on the workpiece (**fig. 6-44**).

6-42: Wedged, through mortise-and-tenon joints on a Shaker-style bench.

6-43: The pattern for routing the mortises shown in **fig. 6-42**. The thickness of the workpiece is enough to support the router.

6-44: With the pattern clamped below the work-piece, a flush-trim bit will rout the mortises.

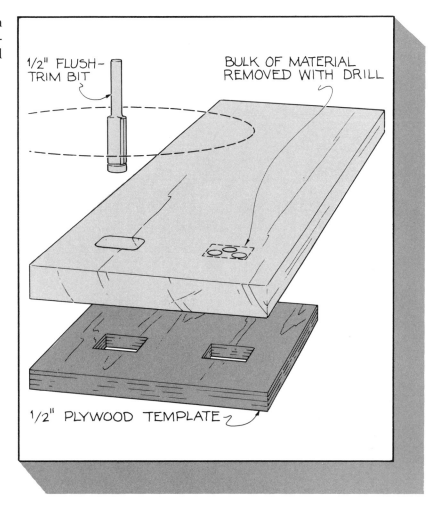

1/2" FLUSH-TRIM BIT

BULK OF MATERIAL REMOVED WITH DRILL

1/2" PLYWOOD TEMPLATE

The Tenon

Tenons, as shown in **figure 6-45,** are most often made on the table saw but can easily be made with a router. One way to think of a tenon is as a piece of wood with rabbets cut on four sides (**fig. 6-46**). Tenons are cut in almost the same way that rabbets are cut (**fig. 6-8**). Using the router table, chuck a ¾-inch (19-mm) straight cutting carbide bit. The height of the bit is such that when the workpiece is run through on each side, the appropriate size tenon remains. Therefore, the wood for your test cut and for all of your pieces must be of the exact same thickness. If the cut is particularly deep, you will probably want to make it in two passes. If the tenons that you are making are very large, waste as much material as you can with a bandsaw, or even a handsaw, before routing (**fig. 6-47**).

The fence setting is such that the distance from the fence to the far, outside edge of the cutter is equal to the length of the tenon. Your first cut will be as shown in **figure 6-48.** Then, using a miter gauge, bring the wood away from the fence and remove the

6-45 (left): Here are four routed tenons. **6-46 (right):** A tenon is really just an endpiece with rabbets cut on all four sides.

6-47: You will make things easier for your router if you use a bandsaw or handsaw first to remove the bulk of the material.

6-48: Cutting a tenon on the router table. The fence setting establishes the length of the tenon. A support piece, as shown in **fig. 6-49**, will prevent chipping.

FENCE MUST BE CLAMPED PARALLEL WITH MITER GAUGE GROOVE

ROUTER TABLE

6-49: Cleaning up the rest of the tenon. Tenons under ¾ inch (19 mm) will be cut in one pass, as with a rabbet.

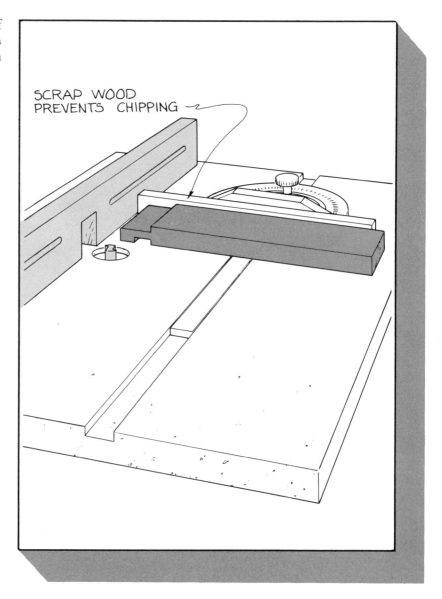

SCRAP WOOD
PREVENTS CHIPPING

remaining material (**fig. 6-49**). The miter gauge, by the way, will give you more control when routing tenons, particularly when routing the short shoulders. Remember, when you use a miter gauge, the fence must be set parallel to the miter-gauge groove. In addition, using a support block behind the workpiece will help prevent chipping (**fig. 6-49**). The short shoulders are cut in quite a bit deeper than the side shoulders; I often cut them with a tenon saw, as shown in **figure 6-50**, rather than rout them.

To make tenons that fit the mortises shown in **figure 6-42,** begin by scribing a line indicating the length of the tenons (**fig. 6-51**). Mark the position of the tenons and cut them with a tenon saw (**fig. 6-52**). Next, bandsaw the bulk of the wood away. Lay a trim guide over the scribed line and rout a clean edge on the outside and between the tenons (**fig. 6-53**). If the tenons are to be shouldered on their faces, you should cut the shoulder before cutting the tenons, using a basic rabbet setup as shown in **figure 6-54**.

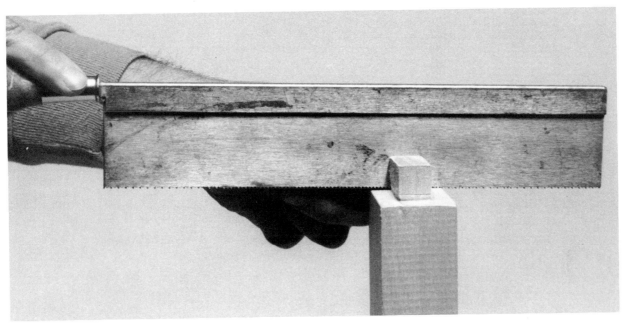

6-50: I usually cut the short shoulders with a handsaw after routing the long shoulders.

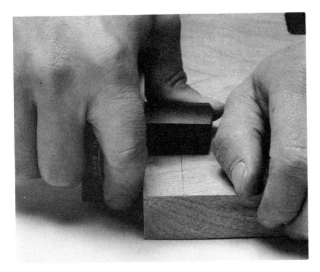 

6-51 (left): Scribing a line to indicate the length of the intended tenons. **6-52 (right):** Using a handsaw to cut the tenons.

6-53: Using a router and trim guide to finish the joint. Note that the trim guide has been placed exactly along the scribed line.

6-54: If you want to shoulder the tenons, do so before the tenons have been cut. Shouldering is making a shallow rabbet cut on each side of the workpiece.

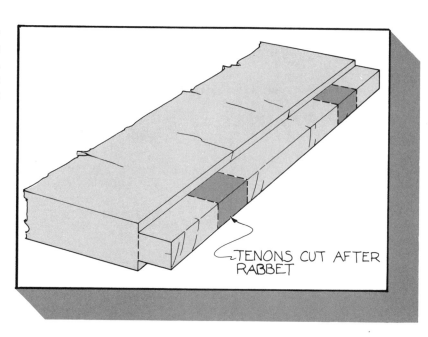

TENONS CUT AFTER RABBET

THE SPLINED MITER JOINT

The splined miter joint is not a joint that one usually associates with the router, perhaps rightly so. In contemporary furniture, where the miter is used so much, the professional relies almost exclusively on the table saw and, to a lesser extent, the radial arm saw, to make miters. But there are a few instances where a routed miter is called for. Certainly those without a power saw will appreciate the following technique.

The easiest way to cut a miter is with a large, 45-degree chamfer bit (**fig. 6-55**). This method is usually reserved for mitering thin stock, ½ inch (12 mm) or less. Set the router-table fence flush with the ball-bearing pilot of the bit. Use several light passes to make the miter, cutting to the exact corner of the wood on the final pass. The mitering will not change the overall dimension of the wood, so all of your pieces should be cut to exact, final dimensions. Chamfer-bit mitering is particularly useful when working with very small pieces, or when stopped miters have to be made.

A second method for mitering is utilizing a 45-degree platform and a moving router. You will first have to build the basic jig shown in **figures 6-56** and **6-57**. Position (glue and nail) the fence so that when the router is moved along the fence the bit routs the

6-55: A 45-degree chamfer bit will cut miters on thin stock. Make the miter in several light passes. Note that the pilot of the bit is flush with the fence.

6-56: Here I am using the mitering platform.

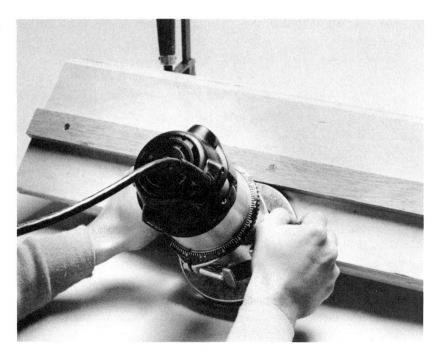

6-57: Dimensions for the mitering platform.

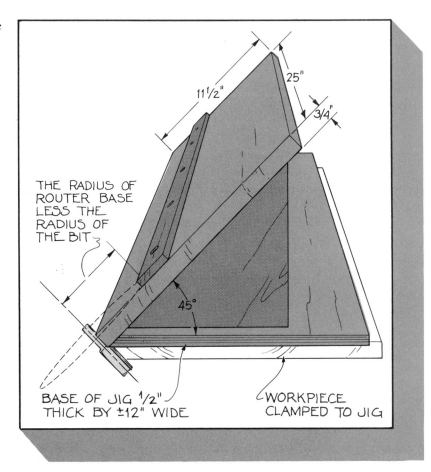

25"

11½"

3/4"

THE RADIUS OF
ROUTER BASE
LESS THE
RADIUS OF
THE BIT

45°

BASE OF JIG ½"
THICK BY ±12" WIDE

WORKPIECE
CLAMPED TO JIG

6-58: To prevent chipout, rout the last inch or so from right to left.

edge cleanly. This method is based on the assumption that you will use the same bit each time with this jig. I would suggest chucking a ½-inch (12-mm) straight cutting carbide bit 1½ inches (38 mm) long and preferably with a ½-inch (12-mm) shank.

You should first rough-cut the miters with a circular saw or saber saw ⅛ inch (3 mm) beyond the point to be mitered. Position the jig so that the bevel on the jig lines up with the miter line, and cut the miter. Take care not to splinter at the end of the cut. I suggest routing the last inch or so in the right-to-left direction, to prevent splintering (**fig. 6-58**).

Adding a spline is easy. Leave the jig in position, chuck an ⅛-inch (3-mm) carbide slotting cutter, and rout the groove (**fig. 6-59**). Depending on the size of the slotting

6-59: Using the miter jig and slotting cutter to route a spline groove.

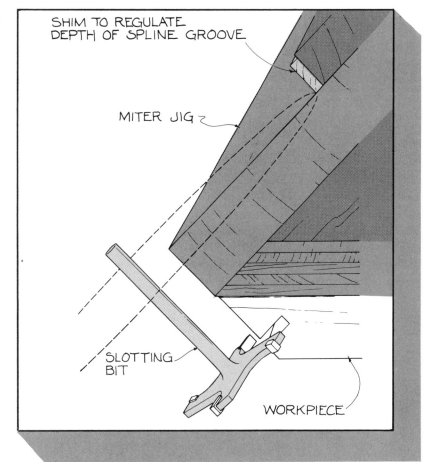

SHIM TO REGULATE
DEPTH OF SPLINE GROOVE

MITER JIG

SLOTTING
BIT

WORKPIECE

cutter, you may have to add another block to the fence to prevent the slotting cutter from cutting in too far. Making a stopped spline (one that does not come through to the end) is easy when you use this technique.

Mitered-Frame Splining

Mitered frames, as shown in **figure 6-60,** often have hidden splines, which are best made on the router table. The spline grooves are usually ¼ inch (6 mm), so you will need to chuck a ¼-inch (6-mm) straight cutting bit. The cut is started by aligning your workpiece with stop block clamped to the fence (**fig. 6-61**). Advance the workpiece forward until it hits the outfeed stop block. As with mortising, you cannot simply turn the workpiece around and cut a spline groove on the other miter. If you did, you would be presenting a different face against the fence on each cut. Cutting the second spline groove is shown in **figure 6-62.**

Decorative splines are often added to mitered frames, especially picture frames (**fig. 6-63**). The most frequently used method for making this spline groove is with the jig shown in **figures 6-64** and **6-65.** An alternative method is to use a clamp-on splining fence and slotting cutter (**fig. 6-66**).

6-60: A hidden spline adds strength to a mitered frame.

6-61: The spline groove is controlled by two stop blocks.

6-62: Cutting the second spline groove.

6-63: Decorative splines add strength to a miter.

6-64: The most frequently used router system for making decorative splines. The jig is passed over a ³⁄₁₆ inch (4.7-mm) straight cutting bit.

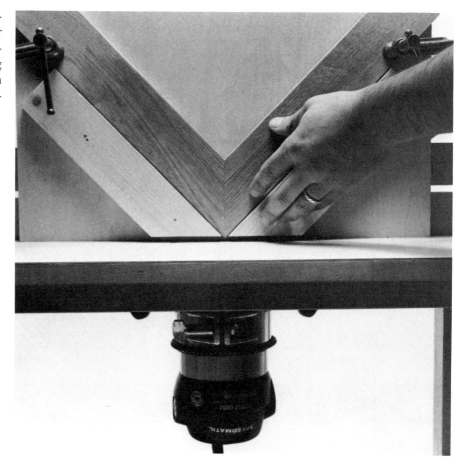

6-65: The same jig can be positioned on the flat and a slotting cutter used instead. You might want to add a piece of scrap wood on the exit side of the miter to prevent chipout.

6-66: Here is a clamp-on splining fence used with a moving router.

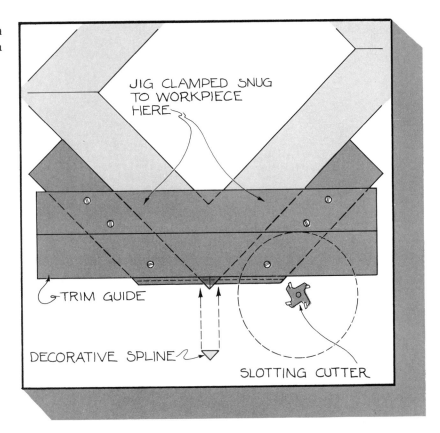

THE LAP JOINT

The lap joint, in its several variations, is used in different types of frame construction. The pieces to be joined are usually the same thickness and width. The router cuts, therefore, are of the same depth and width on each of the mating pieces.

A corner half lap, as shown in **figure 6-67**, is best done on the router table. As with a tenon cut, there is usually a substantial amount of material to be removed, so rough cut the bulk of the material with a bandsaw or handsaw.

A middle half lap, as shown in **figure 6-68**, can be done in the same way, but only if the joint falls near the ends of the boards. If it

does not, you will have to forgo the router table and use a moving router. Clamp your pieces side by side, as many as you can (**fig. 6-69**). Mark the width of the cut and lay a trim guide along one of these lines. Rout across your boards. Reverse the trim guide until all the waste is removed. You can see how important it is that the boards are clamped side by side so that they do not shift while the trim guide is being clamped and unclamped. The tee half lap, as shown in **figure 6-70**, is cut in the same way, even though the lap occurs at different points of the workpiece (**fig. 6-71**).

A dovetail half lap is begun by routing a

6-67: I joined the four corners of my drafting table with a corner half lap.

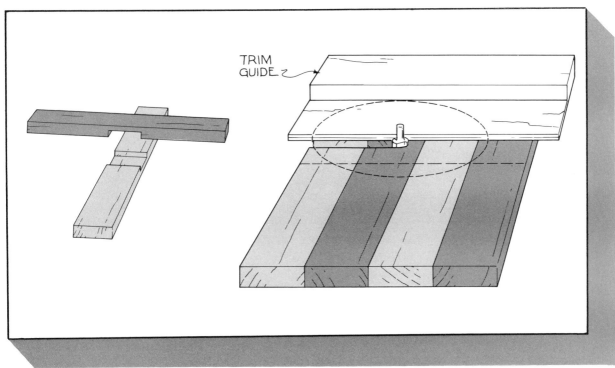

6-68 and **6-69:** Cutting middle half laps by routing the workpieces simultaneously. A trim guide is used for easy positioning.

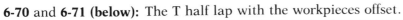

6-70 and **6-71 (below):** The T half lap with the workpieces offset.

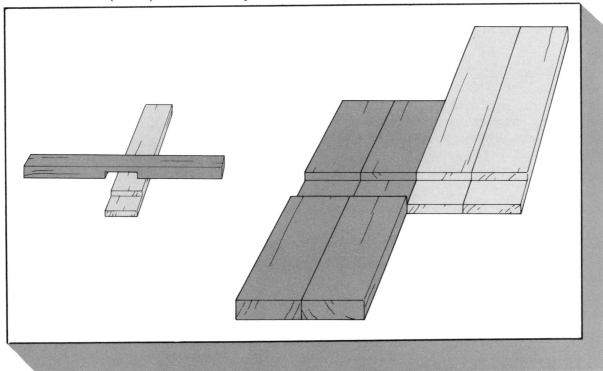

corner half lap. Cut the dovetail itself with a handsaw. If you have many of these joints to cut, make a template of the dovetail from ¼-inch (6-mm) plywood and tack it to the inner edge of the half lap (**fig. 6-72**). Cut the dovetail just to the outside of the template, taking care that the template does not get nicked. Chuck a flush-trimming bit and trim the dovetail carefully to the smooth, final shape. The corners will have to be chiseled square (**fig. 6-73**).

Lay the dovetail on its mating piece and trace its outline. Place a trim guide on the inner edge of this pencil line and rout to a depth equal to the thickness of the dovetail (**fig. 6-74**). Repeat on the other side.

6-72 and **6-73**: Cutting the dovetail of a dovetail half lap with a flush-trim bit and template. The bulk of the wood was removed with a saw.

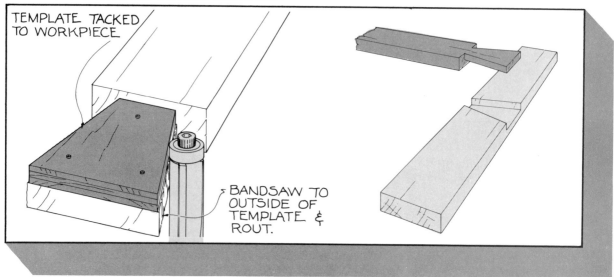

6-74: Cutting the recess for the dovetail half lap. A trim guide is placed along the line of cut.

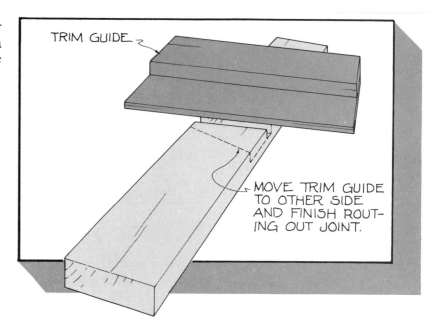

DOVETAILING

The dovetail, a wonderfully strong and attractive corner joint, has for centuries been cut by hand. With one of the several types of dovetail routing fixtures available, however, this joint can be made with a router.

Let us first get the nomenclature straight. Although the dovetail has many structural and aesthetic variations, two major categories of dovetails include the through dovetail and the half-blind dovetail. In a through dovetail, as shown in **figure 6-75,** the joint is visible from both the front and side. This joint is used principally in carcase construction, such as on a blanket chest, but is also found in drawer construction, where the joint is intentionally exposed. In a half-blind dovetail, the joint is visible from only one side, making it a good choice for drawer construction (**fig. 6-76**). The dovetail itself is the flaring member of the joint; the pin is the end grain section that fits between the dovetails. A dovetail that is quite a bit larger than the pin has become the aesthetic standard.

Until recently, the only type of routed dovetail fixture available was one that made half-blind dovetails (**fig. 6-77**). That jig has been sold widely for many years and is the type of fixture most closely associated with routed dovetails. Among the companies that

6-75: A through dovetail. This particular dovetail was made with the Leigh dovetailing jig.

6-76: A half-blind dovetail is usually reserved for drawer construction.

manufacture this jig are Sears, Porter-Cable (formerly Rockwell), Bosch (formerly Stanley), and Black and Decker. (See *Sources of Supply*.) This jig is relatively inexpensive and is used primarily in the making of drawers. A common criticism of this type of dovetailing jig is that the dovetails produced are equal in size and spaced in a fixed way. This gives it a "machined" look, when one expects to see a hand-cut dovetail. In drawer construction, however, a half-blind dovetail has a very definite place.

Your first encounter with one of these jigs can cause a little head-scratching, given the array of adjusting screws, knobs, and stops. Coupled with this is the fact that the boards are routed inside out (**fig. 6-77**). With a little familiarity, however, you will see that the system is all very logical; in time, you will be able to make the various adjustments without much puzzling over them (my criterion for any effective system).

6-77: With this type of dovetailing jig, the pieces to be joined are routed simultaneously. The boards are automatically offset by stop screws and have to be clamped "inside out."

Begin with a test cut, making sure that the test pieces are identical in thickness to the actual workpieces. You will often find that the joint is too tight, too loose, too shallow, or too deep. These problems can be adjusted by regulating the depth of the bit or regulating the workpieces relative to the finger template. Once you have it adjusted, the fixture will enable you to rout consistently clean dovetails. Dovetail fixtures come with explicit directions.

The pieces to be routed are clamped into the jig simultaneously. They are placed inside out and are automatically offset by the stop screws. The finger template is screwed into place. A dovetailing bit and a template guide are inserted into the router. The template guide will "steer" the bit in and around the finger template (**fig. 6-78**).

The beauty of this particular dovetailing system is that the dovetails and pins are routed simultaneously, with the spacing predetermined by the finger template. Because the dovetailed piece is backed by its mating piece, there is virtually no tearout as the cut is made.

Two other dovetail fixtures now available to the woodworker are the Keller Dovetail Template and the Leigh Dovetail Jig. Both of these jigs produce through dovetails with the dovetails sized proportionately larger than the pins, which gives them a traditional, hand-cut look. Considerably more expensive than the jig previously described, these fixtures are designed for small production, or perhaps for the amateur who is not interested in cultivating the skill of hand dovetailing.

The Keller dovetail set, designed by California woodworker David Keller, is designed for carcase dovetailing. It consists of a pair of aluminum templates and a pair of industrial carbide-tipped router bits (**fig. 6-79**). Ball bearings, mounted above the cutting

6-78: A close-up view of the template, around and through which the bushing guide rides.

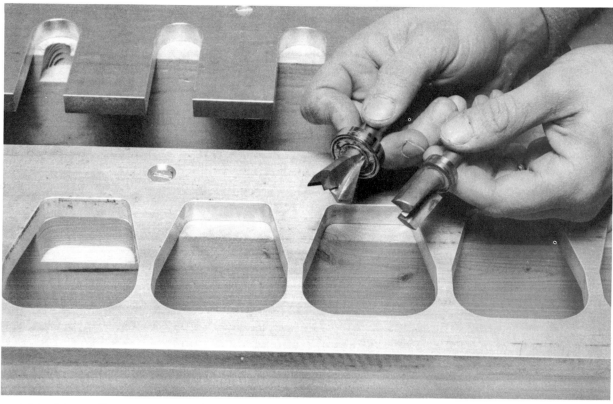

6-79: The Keller jig consists of a pair of industrial router bits and a pair of beautifully machined templates. Note that the ball bearing pilots are placed above the cutters.

6-80: The Keller jig produces a large, well-proportioned dovetail. The pins are 1 inch (2.54 cm) and 3 inches (7.62 cm) on center.

edges of the bits, guide the router along the template. The dovetail bit is ⅞ inch (22 mm) long, has a 19-degree angle, and is used to cut the dovetails. The straight cutting bit is 1 inch (25 mm) long, ¾ inch (19 mm) in diameter, and is used to cut the pins. The pins produced are 1 inch (25 mm) (measured at the widest point) and 3 inches (7.6 cm) on the center (**fig. 6-80**). The templates are 36 inches (91.4 cm). The Keller system is fast, direct, and produces a well-proportioned through dovetail. Its sole limitation is that it produces only one size dovetail. That can be seen as a virtue, however, for it does one thing and does it well. For more information, contact David Keller (see *Sources of Supply*).

The Leigh jig, as shown in **figure 6-81,** was designed by Ken Grisley and also makes through dovetails for both case and drawer construction. Available in both a 12-inch and 21-inch (30.4- and 60.9-cm) size, it is an ingeniously engineered dovetail fixture. The jig is made up of a series of adjustable guide fingers, which slide along on a heavy aluminum extrusion. The fingers are angled on one side and straight on the other. Their adjustability provides for totally flexible dovetailing; that is, the spacing and size of the dovetails can be varied to suit personal taste **(fig. 6-82)**. Both pieces to be dovetailed are clamped to the jig. The dovetail piece is clamped to the side with the straight guide fingers, and the pin piece is attached to the side with the angled fingers. Grisley suggests positioning the angled fingers by eye rather than by measurement. I balked at this at first, but the eye is accurate and the fractionally asymmetrical spacings create the subliminal effect of a hand-cut dovetail.

Once you have got the fingers at the de-sired spacing, lock them in position by tightening the setscrews. A 14-degree dovetail bit cuts the dovetails and either a ⁵⁄₁₆-inch or ½-inch (7.9- or 12-mm) (depending on the thickness of the stock) straight cutting bit cuts the pins. The bits are guided through the template with a template guide.

In making the test cuts, the dovetails are cut first **(fig. 6-83)**. This order is followed because it is in the cutting of the pins that the joint can be adjusted for tightness or looseness by removing or adding paper shims **(fig. 6-84)**. Once you get a successful test cut, it makes no difference whether pins or dovetails are cut first.

My only criticism of the Leigh jig is the problem of grain tearout. This difficulty can be obviated to a large degree by placing scrap wood behind the workpiece. All factors taken into account, the Leigh jig is an excellent system for routing dovetails. It is certainly the most versatile jig. For more information, you should contact Leigh Industries Ltd. (see *Sources of Supply*).

6-81: The Leigh jig has a series of guide fingers that can be adjusted, allowing you to produce dovetails of different sizes.

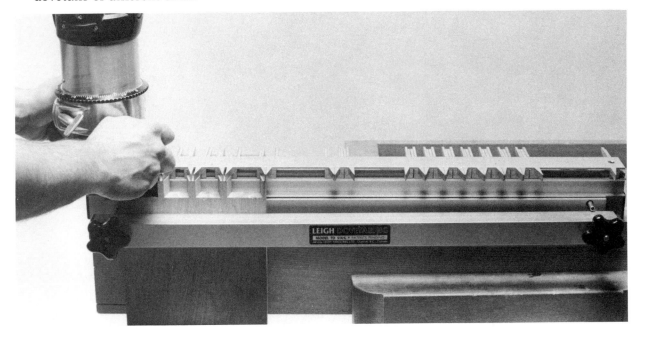

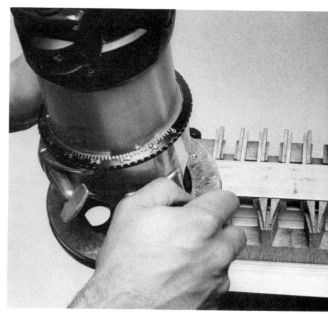

6-82 (left): Note that the top and bottom dovetails are larger than the two inner dovetails. **6-83 (right):** In making test cuts on the Leigh jig, the dovetails (far side) are cut first. When cutting the pins, the tightness of the joint is regulated with shims.

6-84: Cutting the pins.

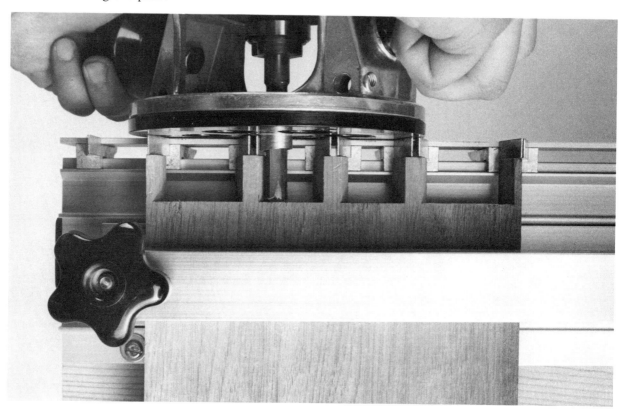

7 Pattern & Template Routing

Template routing and pattern routing are essentially synonymous terms, although template routing suggests to some that a template or bushing guide is used. Pattern and template routing encompasses a wide variety of objectives and techniques, but the common factor is that the router is guided along a designed pattern in order to duplicate it. The patterns may be purchased, as with any of the door decorating kits, staircase templates, and lettering templates, or they may be homemade. In the course of time, a woodworker doing a lot of pattern work will make many more templates than he will buy.

CONTROLLING THE ROUTER

There are three basic systems for controlling or guiding the router against a pattern. In the first, the router base is simply guided along the pattern. This method is seldom used in pattern routing because the large radius of the base limits the pattern to gentle curves. The bit and base, too, may not be perfectly concentric. In routing situations where extreme precision is required, this becomes a serious problem.

In the second system, a template guide rides along the pattern while the bit does the cutting. These guides, as shown in **figure 7-1,** come in a wide variety of sizes, with different inside and outside diameters and in different lengths. Since the template guide fits over the bit and is affixed to the router base, its inside diameter must be fractionally

larger than the bit being used. The length of the guide must also be less than the thickness of the pattern. Template guides are attached to the router base in a number of different ways. Some guides are threaded and screw directly into the router base, some are secured by a retaining ring, and others are held by machine screws. For this reason, template guides are not usually interchangeable from one router to another.

To make a pattern of the correct size, you must first measure the distance from the outer edge of the bit to the outside edge of the template guide (**fig. 7-2**). The template must be made smaller or larger than the final desired shape by this dimension (**fig. 7-3**). See the section "Using a Template and Template Guide," found in Chapter 3.

7-1: Various-size template (bushing) guides.

7-2: Measure the distance from the outside edge of the bit to the outside edge of the template guide.

7-3: The template guide must be made smaller or larger than the final desired shape. See text.

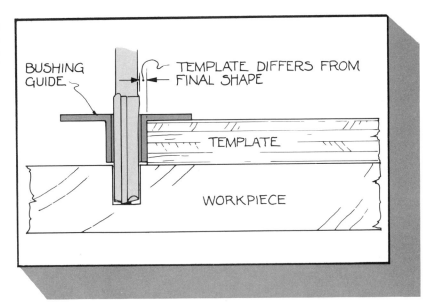

Template guides reduce the problem of nonconcentricity. They are used extensively in decorative line work, irregular grooving, pierced work, and in conjunction with any number of factory-made templates, including staircase, lettering, dovetailing, butt-hinge templates, and door decorating kits.

Figure 7-4 shows how a template guide is used to make a groove for a tambour door.

The third system for controlling the router, which utilizes a flush-trim bit, is the simplest and most direct. The ball-bearing pilot rides along the pattern while the cutting edges of the bit trim the workpiece flush (**fig. 7-5**). Not

7-4: Routing for a tambour door. The template was made fractionally smaller than the final desired tambour groove.

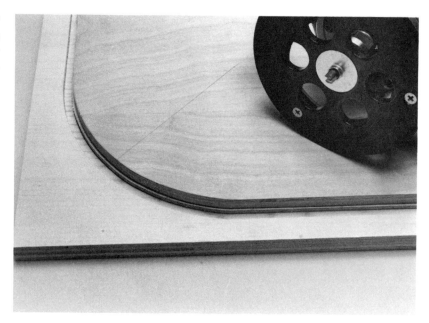

7-5: The ball-bearing pilot rides along the pattern while the bit does the cutting.

only is nonconcentricity no longer a problem, but you can also make your pattern to exact size, unlike templates made to be used with a template guide. This makes it easier to make the pattern and place it on the workpiece. I am amazed that more woodworkers do not take greater advantage of the flush-trim bit.

Standard flush-trim bits usually have cutting diameters of ½ inch (12 mm) and cutting lengths of up to 1½ inches (38 mm). With a little looking, you can find bits with 2-inch (50-mm) cutting lengths. See the Greenlee Tool Division in *Sources of Supply*.

One limitation with the flush-trim-bit system is that, because of the pilot, the bit cannot cut to in-between depths. In other words, you must cut through the workpiece completely. One way around this is to find a bit manufacturer that will outfit you with a bit that has a pilot mounted *above* the cutting edges. This type of bit allows the pilot to ride along the template while the bit cuts to any depth (**fig. 7-6**).

7-6: Making a pattern cut with the ball-bearing pilot mounted above the cutting edges of the bit. Unlike a flush-trim bit, a bit of this type can be used to cut to "in-between" depths.

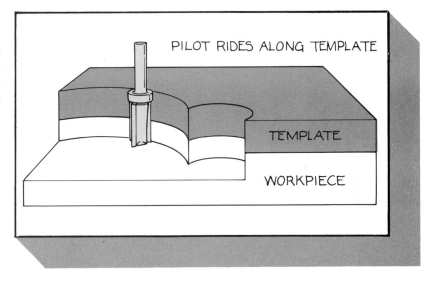

REPETITIVE PATTERN CUTTING

Repetitive pattern cutting refers to reproducing many pieces of the same shape. Rather than cutting and shaping each piece individually, you can make a prototype and use the router to duplicate it.

Making and Attaching the Pattern

Patterns are best made from wood, plastic, aluminum, or masonite. Wood or masonite are the most logical choices because scraps of these materials usually abound in the shop or workplace. Avoid soft woods such as pine. They dent easily, and pressure from a pilot will often leave an imprint. I am partial to ½-inch (12-mm) Finland or Baltic birch, which is a dense, multilayered plywood. Half-inch material is the thickness most commonly used. However, sometimes you may need to use ¼-inch, ⅜-inch or ¾-inch (6-, 9-, or 19-mm) stock.

Make your templates carefully. Bandsaw any curves, but use the router and straightedge or the table saw to cut any straight runs. Curves should be shaped further with rasps, files, and sandpaper. Concentrate on keeping the edges square to the face.

If your pattern requires symmetrical or repetitive shapes, use a sectional template to make the master template (**fig. 7-7**). Using

7-7: Using a sectional template to make a template. The sectional template will ensure that all four corners are identical.

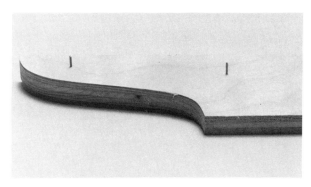

7-8: Using small brads (nails with their heads cut off) to attach a template to a workpiece.

are not practical. Embed small brads (cut off the nail heads on an angle after nailing) or glazier points into the template as shown in **figure 7-8,** then press the workpiece onto these points to hold it in place. I usually reserve this method for times when one side of the wood will not be seen, or will be painted, or when there are many cuts to be made. Attaching handles to a consistently used pattern will make your work easier. **Figures 7-9** and **7-10** show repetitive pattern cutting in action.

a sectional template ensures that the master template will have precisely the same shapes throughout.

The edges of any template should be sanded smooth. Machine oil or a thin coat of wax buffed smooth will provide a silky ride for the template guide or ball-bearing pilot.

Lay the pattern on the workpiece. Carefully trace the outline. Using a bandsaw or saber saw, remove the excess wood just to the outside of this line. The closer you get, the easier it will be to rout. Now, reattach the pattern to the workpiece. This is usually done with clamps, so, although your pattern may consist of one small curve, make it larger to provide room for clamping. If the pieces are to be cut on the router table, then clamps

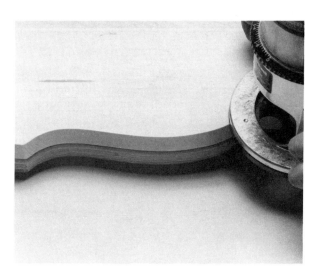

7-10: Repetitive pattern cutting with a moving router and flush-trimming bit.

7-9: Repetitive pattern cutting on the router table. Handles attached to the template make it easier to guide the workpiece.

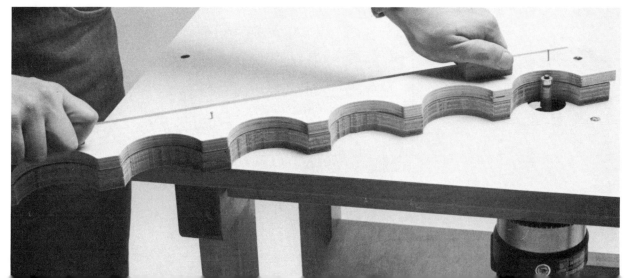

PIERCED WORK

The techniques for pierced work are an extension of the techniques I have been describing, except that you will be working with interior rather than exterior perimeters. Pierced work is most often done with template guides and straight cutting bits, but in some cases you would do just as well with a flush-trim bit. The chief advantage of the former technique is that if a small enough template guide is used (e.g., ¼-inch [6-mm] outside diameter), it can negotiate tighter curves and corners than a flush-trim bit. The

chief advantage of the flush-trim bit is that the pattern can be made to exact size, thereby avoiding the measuring required when using bushing guides.

If you have many pieces to reproduce, you might consider using the jig shown in **figure 7-11.** Position the workpiece in the jig and close the hinged pattern over it. Trace the shapes to be cut. Remove the workpiece, and with a drill and saber saw, remove (waste) as much stock as possible. Reposition the workpiece in the jig and rout.

7-11: A hinged-top fixture for routing pierced work. A template guide rides along the pattern. This concept can also be applied to un-pierced work.

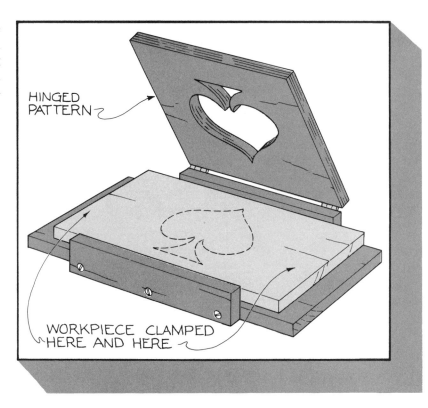

HINGED PATTERN

WORKPIECE CLAMPED HERE AND HERE

DECORATIVE LINE WORK

Decorative line work is routing designed to simulate the carver's parting or veining tool (**fig. 7-12**). For this type of work, you will use either a V-groove or round-nose veining bit in conjunction with a template and template guide (**fig. 7-13**). Decorative line work must be done with great care. One false move and you are in trouble. I have found that large industrial routers are too awkward for this work; lightweight models are better. (An-

other argument for owning more than one router!)

If you are concerned about pulling away from the fence when making decorative line cuts, consider making a template like the one shown in **figure 7-14**. The groove in the template shown was made with a ½-inch (12-mm) router bit. In order to accommodate a template guide with a ½-inch (12-mm) outside diameter, the groove has to be made

7-12: Decorative line work using a round-nose bit, template, and template guide.

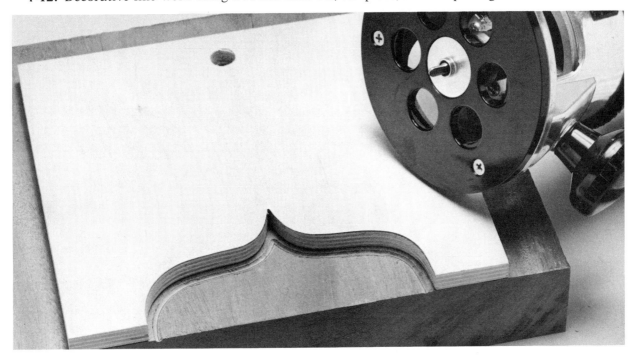

fractionally larger by sanding. With this system, the router cannot go astray.

Door and Drawer Grooving

Commercially available door and drawer decorating kits are available, as one manufacturer says, "to put you in the door business." I recoil a little when the intent of these kit manufacturers is to take plywood doors and simulate the look of a raised panel door. When used on solid door panels and drawer fronts, however, these jigs can be an asset to the woodworker. The jig consists of a metal frame that can be adjusted to any size required. Corner templates can be added for a wide variety of effects, as can a trammel guide to create grooved arcs. An alternative to this jig is to make your own individual templates (**fig. 7-15**).

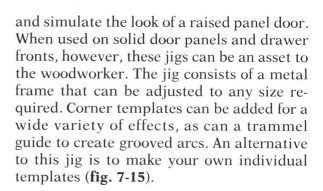

7-15: Design your own templates for adding decorative touches to doors and drawers.

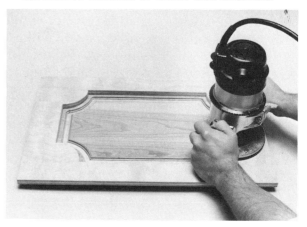

7-13: The V-groove bit can simulate the effect of a carver's parting tool.

7-14: This type of template ensures that the router cannot go astray. The bushing guide is the same size as the groove.

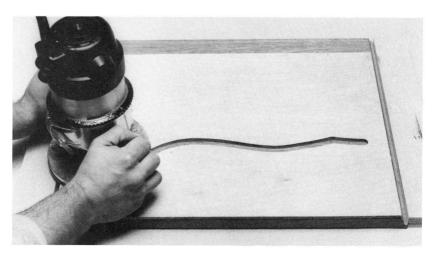

8 Routing for Hardware

The installation of hardware is an important aspect of woodworking. Whether it is the mortising of a hinge or the recessing of campaign hardware, the router is the woodworker's choice for producing a crisp and clean fit. The router has the unique ability to cut to any depth and to start and stop at precise points. It can cut along the edge of a door or in the middle of a drawer face. When used with templates, a router makes exact fits easy. In addition, you can create your own wooden hardware, using a router.

INSTALLING HINGES

I have selected three types of hinges for discussion—the butt hinge, the piano hinge, and the pivot hinge. Along with concealed European hinges, which are drilled rather than routed, these three hinges represent the most common and important hinging systems used in cabinetmaking.

Properly installed hinges are recessed into a notch sometimes called a *gain* but more often called a *mortise*. Specifically, one leaf of the hinge is mortised into the door and the other into a cabinet. This is called *single mortising*. In some cases, cabinetmakers elect to *double mortise* the hinges into the door; that is, they set the full thickness of the hinge into the door and screw the hinge to the cabinet without mortising it (**fig. 8-1**). This practice would never be acceptable when hanging a house door but is sometimes employed on even fine furniture when the weight of a door is insignificant.

Butt Hinges

The butt hinge (also called the barrel hinge) is usually single mortised into both the door and the cabinet. There are probably a dozen ways to rout a butt-hinge mortise. First determine the correct placement of the hinge on the door. On an overlay door, as shown in **figure 8-2**, the hinge is mortised into the back of the door and the forward edge of the cabinet. In a flush mount, as shown in **figure 8-3**, the door is fitted within the opening of the cabinet and the hinges are mounted on the side of the door and the inside of the cabinet.

When laying out a mortise for a butt hinge, you must determine how deep the cut into the door should be. For house doors, each mortise is cut to a depth equal to the thickness of a leaf. Because the barrel of the hinge is thicker than the thickness of the two leaves, an even space of approximately ⅛

8-1: The leaf on the right has been double mortised into the door, while the leaf on the left is simply screwed to the cabinet.

8-2 (left): An overlay door. This butt hinge has been single mortised into the door and the cabinet.
8-3 (right): In a flush mount, the door sits within the frame of the cabinet.

inch (3 mm) is created between the door and the jamb. On cabinet doors, however, flushing the leaves creates too large a gap between the door and cabinet. Here, you would have to rout deeper than the thickness of one leaf, approximately $1/32$ inch (.79 mm) less than half the thickness of the barrel. This would create a $1/16$-inch (1.5-mm) space between the door and cabinet.

The second question is how far in the mortise should be cut. The best way to determine this is by laying the hinge, with the barrel down, against a piece of wood (**fig. 8-4**). Mark a line against the back of the hinge. This line shows how far the mortise should be cut-in. Mark the length of the hinge as well.

The door, which is usually mortised first, is routed in one of two positions, as shown in **figures 8-5** and **8-6.** Routing the door on

8-4: To determine how far in a hinge mortise should be cut, lay the barrel of the hinge against the workpiece and mark with a pencil. This door will overlay the side of the cabinet.

 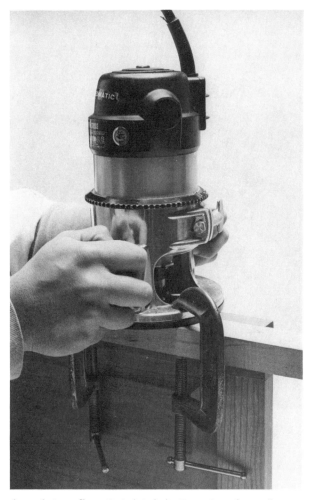

8-5 (left): Routing for a hinge mortise with the door lying flat. **8-6 (right):** Routing for a hinge mortise with the door on edge.

edge, as shown in **figure 8-6,** means that if you want to rout the mortise in the cabinet with the same setting, the router must be positioned on the flat inside edge of the cabinet. This is not always possible, for the router cannot get close enough into the corner. In this case, you would rout the *door* on the flat, as shown in **figure 8-5,** and the *cabinet* on the edge (**fig. 8-7**). Either way, rout from line to line by eye. Corners will have to be chiseled square.

If you have many hinges to mortise, or if you tend to use the same style and size of

hinge over and over, it would make sense to make a hinge-mortising jig. **Figure 8-8** shows one such jig, which utilizes a template guide. A variation of this jig, in which the router base itself rides against three controlling fences, is shown in **figure 8-9.**

For large house doors, carpenters often use a commercially made butt-hinge template (**fig. 8-10**). It is sold as a kit and includes the appropriate template guide. Completely adjustable, this template quickly locates the hinge positions on both the door and the jamb for a perfect match.

8-7: Routing the cabinet for a mortise.

8-8: A homemade hinge-mortising jig in which a template guide rides along the pattern. Corners will have to be chiseled square.

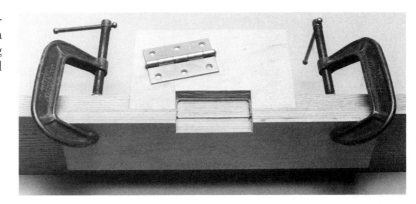

8-9: The dropleaf hinge shown here has leaves of unequal length. I built this template with a removable piece along the back fence so that the same jig could be used to rout both mortises.

8-10: Because I had many of these solid oak house doors to hang, I used this Stanley butt-hinge template.

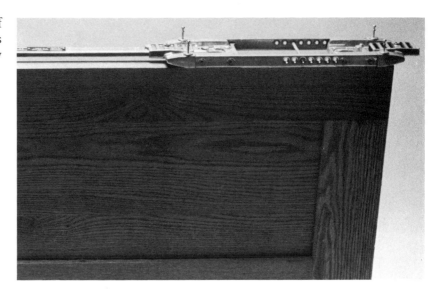

When mortising for butt hinges, keep the following thoughts in mind:

- Always make a test mockup first, to be sure that the depth settings will allow a proper space to exist between the door and the cabinet.
- Mortises are sometimes routed freehand, particularly when there are just a few.
- Mortising bits are designed for fast chip removal, but are hardly necessary for general cabinet work. I use either a ½-inch or ¾-inch (12- or 19-mm) straight cutting bit.
- If you have trouble balancing your router on the edge of a door or cabinet, clamp on an auxiliary support piece (**fig. 8-11**).

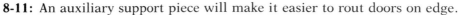

8-11: An auxiliary support piece will make it easier to rout doors on edge.

Piano Hinges

Piano hinges are used in lieu of butt hinges when extra hinge strength is required. Mortising for a piano hinge is really just an extension of the mortising that was done for the butt hinge (**fig. 8-12**). I often prefer to

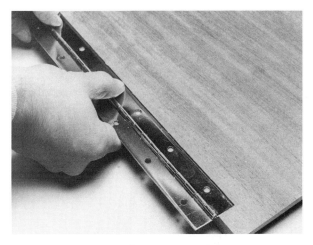

8-12: Mortising for a piano hinge is no different from mortising for a butt hinge, except that the piano-hinge mortise is longer.

double mortise the hinge into a door, dropleaf, or chest lid, and simply screw it into the cabinet. On dropleaves and lid tops, piano hinges are often routed so that they stop short of the ends.

Pivot Hinges

The pivot hinge, as shown in **figure 8-13,** is often the cabinetmaker's choice, not only

8-13: The pivot hinge is inconspicuous and can be used in many styles of furniture.

because it is clean and inconspicuous, but also because it can be used in almost any style of furniture. The process for installing pivot hinges properly may seem a bit complicated. You may want to practice before going on to the real thing.

Although pivot hinges are available in different sizes, I usually use a $5/16$ inch (7.9-mm) hinge on $3/4$-inch and $7/8$-inch (19- and 22-mm) thick doors. The $5/16$ inch (7.9 mm) refers to the width of the leaves. Pivot hinges are typically double mortised into both the top and bottom of the door. Cut the door $1/16$ inch (1.5 mm) shorter than the opening. Using a $5/16$-inch (7.9-mm) straight cutting bit, rout the top of the door first. Set the depth of cut to equal the full thickness of the hinge. Position your router-table fence so that the groove will be in the exact center of the door. The length of the groove is equal to the full length of the hinge. Set a stop block to control the length of this cut (**fig. 8-14**). Make a test cut. Is the depth of the groove equal to the full thickness of the hinge? Is the groove exactly in the center of the door? Is the length of the groove correct? Make any adjustments.

The mortise for the bottom of the door is the same as the one for the top of the door, except that it is ½₂ inch (.79 mm) shallower. This raises the door off the bottom of the cabinet.

Once this first groove has been cut on both ends of the door, you will see that the hinge fits in snugly but cannot open (**fig. 8-15**). Next, make a secondary cut on the inner edge of the door, as shown in **figure 8-16,** removing slightly more than the thickness of one leaf—approximately ⅟₁₆ inch (1.5 mm). This permits the hinge to swing open (**fig. 8-17**).

The final cut is just a notch cut for the protruding part of the hinge. Its depth is equal to the depth of the first groove. I usually chisel this out by hand (**fig. 8-18**).

Allowing the hinge to hang ½₂ inch (.79 mm) over the edge of the door creates space between door and cabinet wall. The hinge is then screwed to the cabinet (**fig. 8-19**).

8-14: With the door on edge, a groove is routed on both the top and bottom of the door. A stop block controls the length of the groove, which is equal to the length of the hinge.

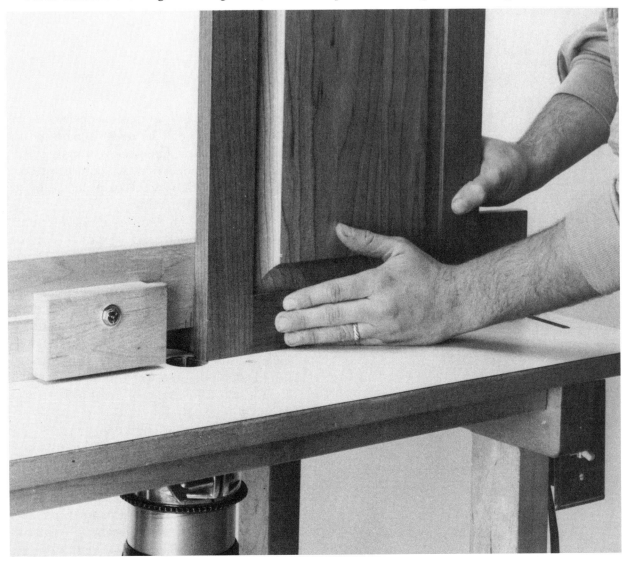

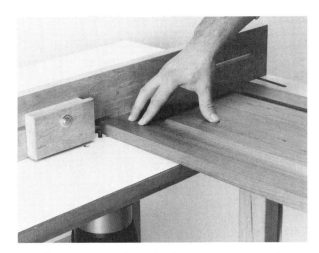

8-15 (left): The hinge fits in but cannot open. A secondary cut to allow the hinge to swing will be made next. **8-16 (right):** The second cut is made with the back of the door facing the router table.

8-17 (left): The first groove, in the center of the door, is equal to the full thickness of the hinge. The second cut, equal to slightly more than the thickness of one leaf, allows the hinge to swing freely. **8-18 (right):** The final cut is a chisel notch for the part of the bit that protrudes forward.

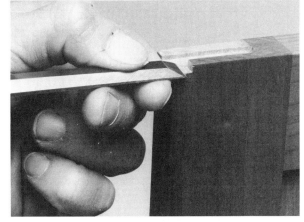

8-19: Screwing the pivot hinge to the cabinet.

DECORATIVE HARDWARE

Decorative hardware, such as handles and pulls, is often recessed into the faces of drawers and doors. I like to do this type of routing in as controlled a way as possible. Few mistakes will raise the hair on a cabinetmaker's neck more than an errant cut on a fitted door or drawer. To this end, I make a jig that controls the cut to the exact dimensions I want (**fig. 8-20**). To rout a pull in a drawer face, begin by cutting a piece of ¼-inch (6-mm) plywood to the size of the drawer. On this piece, lay out the position of the hardware (**fig. 8-21**). Insert a ½-inch (12-mm) straight cutting bit and measure the distance from the outer edge of the bit to the edge of the router base. Carefully transfer this mea-

surement to position the four perimeter fences on the jig. Plunge the cutter through the ¼-inch (6-mm) plywood and rout against the fences. Check to see if the hardware fits into this recess (**fig. 8-22**). If it does, position the jig on the drawer face, set the bit to the proper depth, and rout.

Some hardware, such as the campaign pull, as shown in **figure 8-23**, requires two recesses—one inner, deep cut, and an outer, shallow recess for the brass plate. Make your jig for the outer, shallow recess. By adding additional fences to the jig and lowering the depth of cut, the inner recess can also be routed.

Certain types of decorative hardware, such

8-20: A typical jig for recessing a piece of decorative hardware.

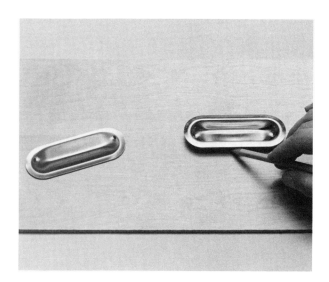 

8-21 (left): Marking the hardware on a piece of ¼-inch (6-mm) plywood. **8-22 (right):** Making sure the hardware fits snugly in the recess.

8-23: This campaign pull requires two recesses. The rounded corners are then chiseled by hand.

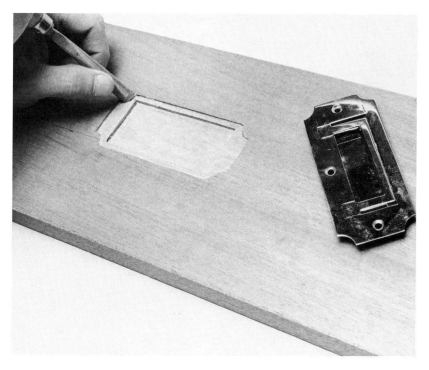

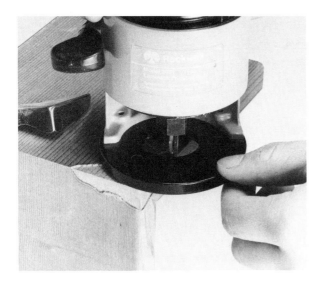

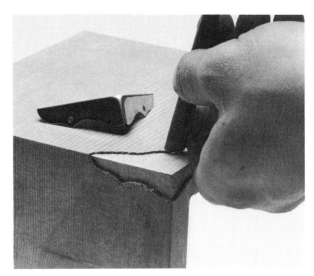

8-24 (left): The recess for this campaign chest corner was routed freehand with a laminate trimmer.
8-25 (right): Finishing the recess with carving tools.

as campaign chest corners, as shown in **figure 8-24,** are best routed freehand. Rout to within ⅟₁₆ inch (1.5 mm) of the line and finish with carving tools (**fig. 8-25**).

Making Your Own Hardware

Certain pieces of furniture suggest the use of wooden hardware. There is very little commercially available wooden hardware, however, and even if you are lucky enough to find a style you like, it will rarely be available in the species of wood you happen to be working with. The only option left is to make your own. **Figure 8-26** shows various examples of routed door and drawer pulls. Your only limitation is your own imagination.

Wooden hardware is usually shaped and routed first and then recessed into the wood-

8-26: Different styles of pulls and handles that can be made with the router.

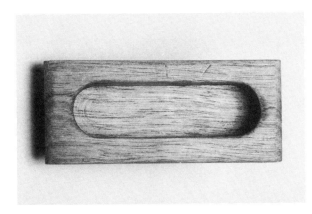 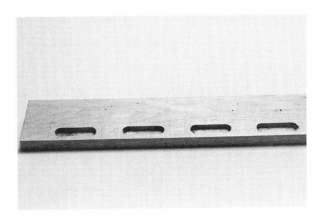

8-27 (left): A router-made sliding-door pull. **8-28 (right):** The grooves for the pulls are routed first, in a series on a wide board.

work. Let me emphasize that this type of routing can be extremely dangerous. It often requires that your fingers come very close to the bits while you are working with very small pieces. A little forethought can significantly lessen the danger. Take, for example, the pull shown in **figure 8-27**. Rather than cut small blanks to rout, it would make more sense to cut the grooves in a series on a large board (**fig. 8-28**). This piece should then be

cut into a long, narrow strip. Rabbets are then cut along the length of the strip's sides, and a rounding-over bit is used to soften the top edges (**fig. 8-29**). After crosscutting the pulls to length, all that remains to be done is rabbeting and shaping the ends. **Figure 8-30** shows one way of safely supporting a small piece when shaping its ends.

Wooden hardware, of course, does not have to be an added piece, but can be routed

8-29: After the board was cut into a narrow strip, a rounding-over bit was used to soften the edges. Now, the rabbets are cut.

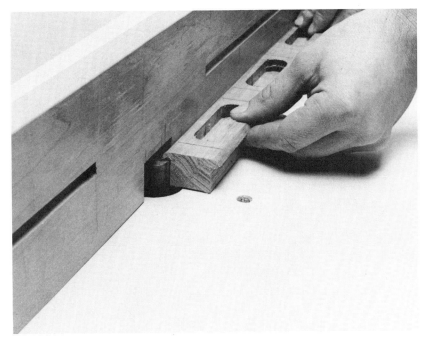

right into a door or drawer. **Figure 8-31** shows a round finger grip that was cut with a template and template guide. The bottoms of grooves will often look better if rounded somewhat. For this type of grooving, use either a round-nose or core-box bit.

8-30: Routing small pieces can be particularly dangerous. This small workpiece was wedged into a larger piece, which is advanced safely over the bit.

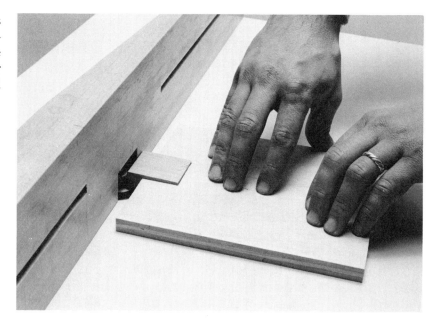

8-31: This round finger grip was routed with a template and template (bushing) guide. A rounding-over bit will be used next to soften the sharp edges.

9 Laminate Trimming

Plastic laminates, such as Formica, are frequently used in contemporary furniture making, and the professional woodworker works with laminates often. "Mica," as it is commonly called, is available in two thicknesses—a vertical grade, approximately $\frac{1}{32}$-inch (.79 mm) thick, used on door and drawer faces, and a horizontal grade, approximately $\frac{1}{16}$-inch (1.5 mm) thick, used on flat surfaces such as countertops. The laminate, which is always cut larger than the wood to which it is applied, is trimmed with either the router or its lightweight cousin, the laminate trimmer (**fig. 9-1**).

In this chapter, I will cover the tools, the bits, and then the techniques for applying laminates and then trimming them precisely.

THE TOOLS

Any router can be used to trim plastic laminates, but a laminate trimmer is a necessity for someone who does this type of work often. The chief advantage of a laminate trimmer is that it is lightweight. It can be held easily with one hand, if need be. Also, it can be used comfortably in a vertical position and will easily balance on the edge of a ¾-inch (19-mm) door. A router, even a small one, is much more difficult to control in these situations.

Laminate trimmers are equipped with edge guides designed to control the amount of bit exposed (**fig. 9-2**). If you try to trim the

9-1 (right): This is a laminate trimmer. **9-2 (below):** The edge guide of a laminate trimmer here controls the amount of cut.

laminate perfectly flush, you run the risk of cutting into the surface against which the edge guide is cutting. The edge guide allows you to adjust the cut so that the laminate is trimmed fractionally oversize, maybe ¹⁄₆₄ inch (.38 mm).

Two variations of the basic laminate trimmer are the offset laminate trimmer and the tilt-base laminate trimmer. In an offset laminate trimmer, as shown in **figure 9-3,** the bit is offset from the shaft of the motor so that it is positioned near the edge of the base. This allows the bit to cut into tight corners, something impossible to manage with a conventional laminate trimmer. Offset laminate trimmers are designed primarily for the professional who does a lot of on-site mica work. The tilt-base trimmer, as shown in **figure 9-4,** is indispensable in a shop such as mine, where I am often requested to build oddly shaped mica-covered pieces. Before I owned a tilt-base trimmer, I used my con-

9-3: The offset laminate trimmer, with its triangular base and offset bit, allows you to rout into corners that would be impossible to reach with a conventional laminate trimmer.

9-4: The tilt-base laminate trimmer has an adjustable base that conforms to any odd-shaped angle.

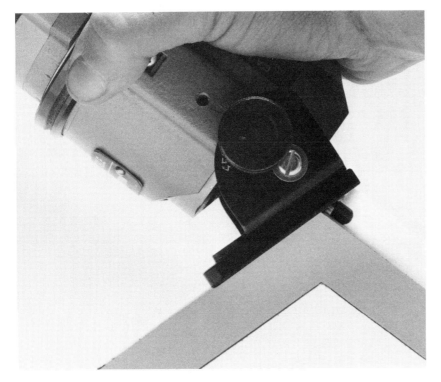

ventional laminate trimmer almost free-hand, leaving a generous overhang and finishing it with a file.

Most router manufacturers offer an edge-guide attachment which, in a certain sense, converts a router to a laminate trimmer (**fig. 9-5**). This edge guide is attached to a sub base that replaces the router's sub base. It has a vernier adjustment to control the amount of cut. An edge-guide attachment cannot, however, solve the problem of using a heavy-weight router.

Using eye goggles is a must when trimming laminate. Your router will spray a stream of hard plastic chips, and it is almost impossible to prevent them from getting in your eyes. Do not forget this precaution.

9-5: This edge guide attachment converts a conventional router into a laminate trimmer. It does not, however, result in a lightweight trimmer, as desired.

THE BITS

There are many types of laminate trimming bits. The one most commonly used is the flush-trim bit complete with ball-bearing pilot (**fig. 9-6**). The ball bearing and bit are offset between $5/1000$ inch and $1/100$ inch (.013 and .025 mm) so that the bit does not cut into the surface against which it rides. This type of bit is used with a router or laminate trimmer but without an edge guide. The bevel-trimming bit, available with bevels of between 10 degrees and 25 degrees, is used to chamfer the sharp, flush-trimmed edges (**fig. 9-7**).

Trim and bevel bits are also available without pilots, in which case they are used with edge guides. Combination trim and bevel bits are also available (**fig. 9-8**). By regulating the depth of cut you can either bevel or trim with the same bit. Two other types of bit of interest here are the flush trimmer and bevel trimmer in solid carbide with a built-in pilot (**fig. 9-9**).

9-6: The flush-trim bit trims the overhanging laminate.

9-7: The bevel-trimming bit flushes and bevels at the same time.

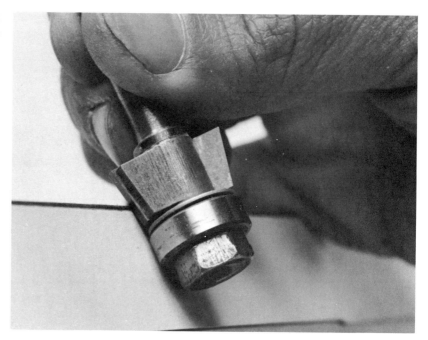

9-8 (above): A combination trim and bevel bit made of solid carbide. Whether it trims or bevels is determined by the relative position of the base. **9-9 (right):** A flush-trim bit in solid carbide with built-in pilot.

APPLYING AND TRIMMING LAMINATES

Although this chapter is about trimming laminates, let me outline a few of the techniques for applying this material. Narrow strips of mica are laid down by hand, as shown in **figure 9-10,** with one hand feeling that the strip is hanging over both sides evenly and the other hand slowly lowering the mica. Lay the mica carefully, for contact cement is an unforgiving fixative. If the strip is angled even a few degrees one way or the other, it will end up misaligned at the far end of the workpiece. Trimming the narrow edges is shown in **figure 9-11. Figure 9-12** shows the correct method for positioning a large piece of plastic laminate. Clean sticks separate the laminate from the workpiece and are removed one at a time as the laminate is pressed into position. **Figure 9-13** shows the accepted sequence for putting mica on a door. Once the laminate has been

9-10: Laying a narrow strip of plastic laminate ("mica"). One hand feels that the laminate overhangs evenly while the other hand lowers it into position.

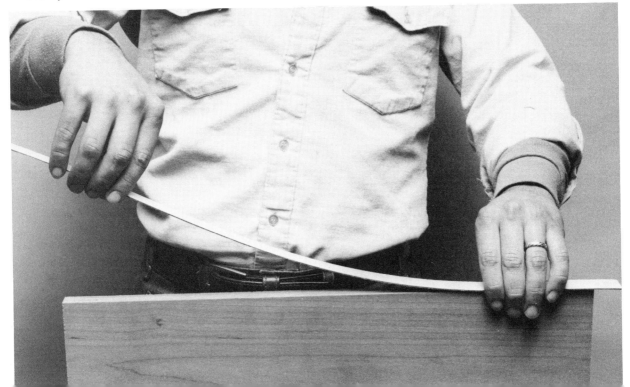

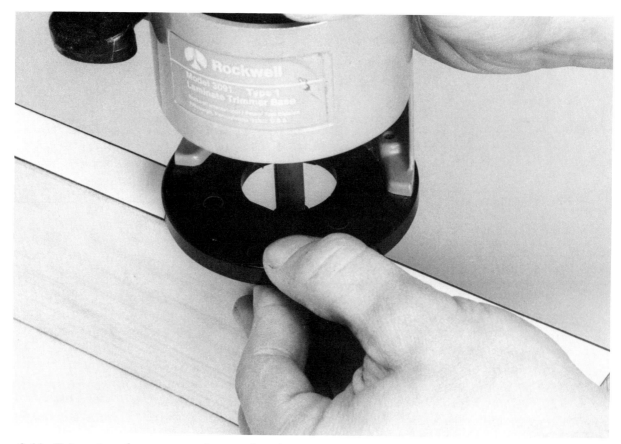

9-11: Trimming the narrow edges with a laminate trimmer. You can place your thumb on the router base to increase stability, but do this with great care since the bit is quite close.

9-12: Large pieces of mica are laid on clean sticks. The sticks, which will not adhere to either piece, allow you to ensure that the mica is overhanging the workpiece on all four sides.

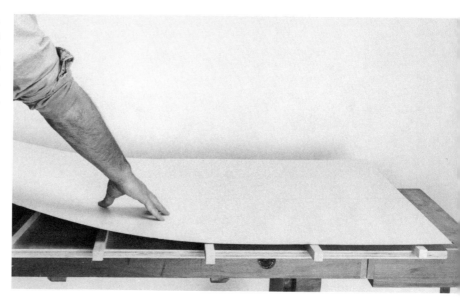

9-13: The correct sequence for covering a door with a plastic laminate.

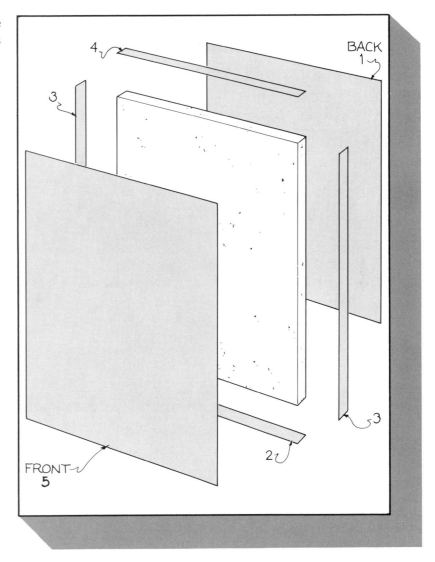

laid down, give it a light pressing with a roller (**fig. 9-14**). Be careful: too much pressure along the edges could result in breakage. When trimming large, horizontal surfaces, I usually keep it simple and insert a standard flush-trim bit with ball-bearing pilot into my 1½-h.p. router. I find that most laminate trimmers labor a little when cutting long runs of horizontal grade micas. I recommend using different parts of the bit (raising or lowering the router base) to ensure even wear. For repetitive trimming, such as on door or drawer faces, I set the same router and bit in the router table (**fig. 9-15**). This is

9-14: Pressing the mica with a J roller.

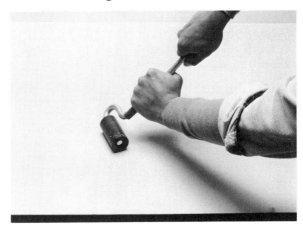

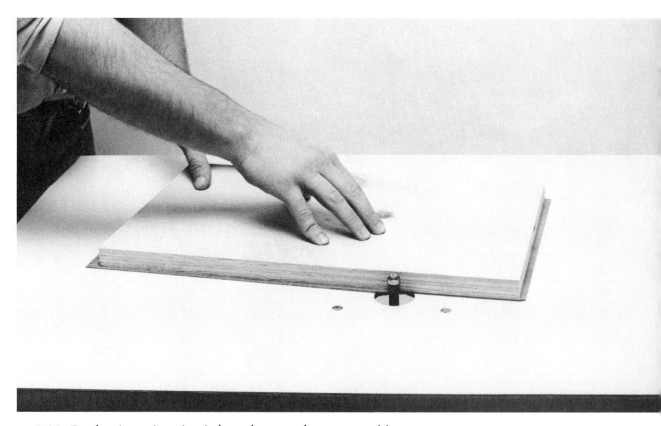

9-15: Production trimming is best done on the router table.

9-16: A quick sanding gives the mica edges a desired smooth feel.

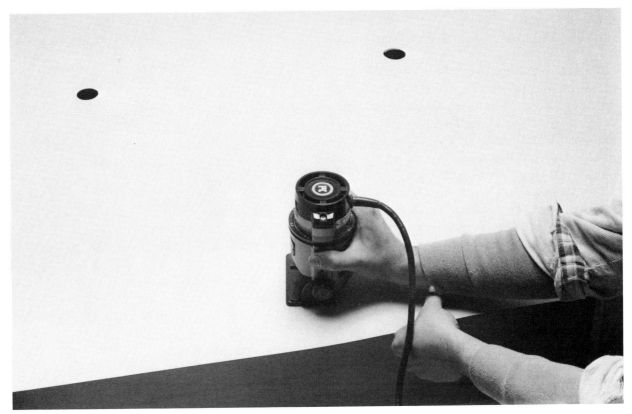

9-17: The best way to lay mica on the face of a cabinet is with a whole piece of material. Predrilled holes give the laminate trimmer a place to start. What you lose in material, you make up in labor. This approach certainly makes for a cleaner face.

certainly the fastest way. Once the initial trimming has been done, I give the surface a more thorough pressing with the roller. I follow this with a pass of a bevel bit. Finally, I soften the edge with 150 sandpaper (**fig. 9-16**).

When putting mica on the faces of cabinets, I sometimes lay a whole piece of vertical grade material over the entire face (**fig. 9-17**). This permits a faster and cleaner application and makes the seamless forward edges of the cabinet look crisp (**fig. 9-18**).

9-18: This shows the seamless corner of a cabinet. Here the inside corner will now be squared with a mica file.

Predrilled holes allow entrance of the trim bit, although you can also purchase a trim bit with a drill point. When routing these edges, as well as the edges of doors and drawers, use a delicate touch. Concentrate on not letting the router tip. A commonly used, but potentially dangerous, technique (if you are not paying close attention) is to place your thumb and fingers around the router base to increase its stability (**fig. 9-11**). A steady hand is also needed when routing the edges of a round surface, because the router has no flat surface to rest upon (**fig. 9-19**).

Working with plastic laminates is not difficult—you will discover many little tricks as you go along. If you have never worked with this material before, you might consider finding someone who frequently does. A few hours of observation will go a long way.

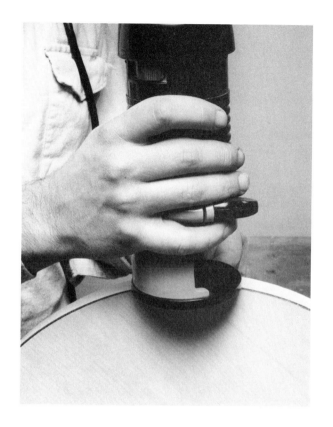

9-19: Routing a round surface requires a steady hand, since the router, as shown, has little bearing surface.

10 Routing Round

The router is not limited to use on flat stock or square wood. With a little ingenuity, it can be set up to cut and shape circles and perform a variety of tasks on round stock.

For the routing of circles, I will show how to make and use a simple jig that can also be used to rout arcs and curved grooves. Round stock, such as a cylinder, is challenging because the router can't be rested on the surface you will be cutting. Nonetheless, accurate routing on the round is feasible. In fact you can do fluting as well as dovetailing. However, owing to the shape of the dovetail bit, the dovetail must be cut in one pass and can't be lowered in progressive stages, as you can when fluting. Since this one-pass requirement for dovetails can put a strain on the router, it is usually advisable to remove some of the wood by doing preliminary drilling.

CIRCLES

There are few things as pleasing as a perfect circle. There are many ways to rout a circle, but I will start with the one method I have been using ever since I started woodworking (**fig. 10-1**). To make a jig, cut a piece of ⅜-inch (9-mm) plywood to a width equal to or slightly greater than the diameter of your router base. Its length should be a little larger than the radii you expect to be cutting. If, for example, you will be making circles of up to 48 inches (1.2 m) in diameter, cut your plywood to 30 inches (76 cm). I have two of these jigs, one 18 inches (45.7 cm) and the other 36 inches (91.4 cm). The smallest circle you can cut using this system is approximately 6 inches (15.2 cm) in diameter. Remove the sub base of your router and screw the plywood in its place. You will have to countersink fairly deep in order to use the same machine screws.

Insert a straight cutting bit. Any size will do, so long as the cutting edge is longer than

10-1: The basic circle-cutting jig was used to rout this round table top. The top was already in two halves prior to routing. Blocks of wood screwed to the underside of each piece held them together while the circle was being cut.

the thickness of the workpiece. Determine the radius of the circle you want to make. Turn the router upside down and transfer this measurement from the outside edge of the router bit to some arbitrary point along the jig (**fig. 10-2**). Drive a 4d finishing nail through at this point. Understand that it is not necessary that this hole be centered on the jig. Pull the nail and renail it from the other side of the jig.

After you find the center of your workpiece, there are two ways to proceed. In the first approach, you would mark the circle with a homemade trammel, as shown in **figure 10-3**, then bandsaw just to the outside of this line. Attach the circle-cutting jig by nailing

10-2 (above): Transferring the desired radius to the circle-cutting jig. The placement of the nail in the jig need not be centered. **10-3 (below):** Marking the circle with a homemade trammel.

at the center of the circle (**fig. 10-4**). For those of you who recoil at the idea of putting a nail in your workpiece, relax. All of this is done on the underside of the circle—that is, the "B" side. Be careful not to drive in the nail too deep. Before starting the router, lift it high so that the bit is not in contact with the wood. Now start it, plunge, and rout in a counterclockwise direction (**fig. 10-5**). There will be sections, however, in which the grain will dictate that you rout clockwise to prevent tearout (**fig. 5-9**).

In the second method, begin by nailing the jig in place. Set the depth of cut to ¼ inch (6 mm). Plunge and rout. Lower the bit another ¼ inch (6 mm) and repeat. With this

10-4 (above): Attaching the circle-cutting jig. **10-5 (below):** Routing the circle. Make sure the bit is not touching the workpiece when you start the motor.

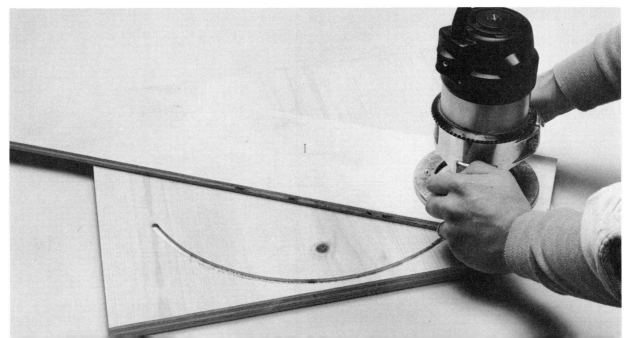

10-6 (left): The circle-cutting jig is just as appropriate when it is the hole and not the circle that you want—for example, a stereo speaker opening. **10-7 (right):** Indexing the circle-cutting jig is done on the assumption that you use the same size bit each time.

technique, the router has to work a little harder, but you are spared the effort of layout and bandsawing. This is also a good system to use when it is the hole and not the circle that you want (**fig. 10-6**).

To index your circle-cutting jig, write next to the nail hole the size radius produced. Another way is to mark a series of measurements, say ½ inch (12 mm) apart, for quick reference (**fig. 10-7**). It is assumed that you will use the same size bit each time.

Router manufacturers do make circle-cutting attachments. Their main advantage is the ease of adjustment they allow, but they require that you drill a ¼-inch to ⅜-inch (6- to 9-mm) hole in the workpiece. You can make the circle larger or smaller by simply loosening and tightening a screw.

Repetitive Circle Cutting

If you have many circles of the same size to cut, first rout a master circle from ½-inch

(12-mm) plywood. Place this on top of the piece to be cut. Pencil-in the shape, then bandsaw. Reattach the blank to the master and clean up with a flush-trim bit (**fig. 10-8**). This system is also good when you do not want a nail hole on either side of your workpiece, because the master and workpiece can be clamped together. You will, however, have to stop and shift clamps as you rout your way around the circle.

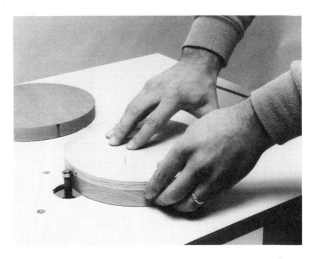

10-8: Repetitive circle cutting is easy with a template and a flush-trim bit.

ARCS

Arcs are routed using the basic circle-cutting jig. If you do not want to put a nail hole on either side of the workpiece, clamp a board to the workpiece and nail into that instead (**fig. 10-9**). Then, of course, you will have to tack a block of equal thickness under the router. Another trick when cutting arcs (and circles for that matter) is to make the first pass a very shallow one. Now that the curve is laid out for you, trim to just the outside of it with a saber saw (**fig. 10-10**). Now rout to the full depth.

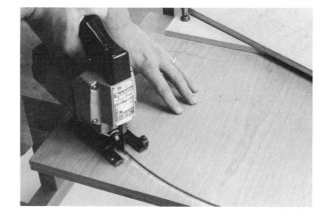

10-9 (below): Routing an arc. The circle-cutting jig is nailed to a block that has been clamped to the workpiece. A block of equal thickness has to be tacked under the router. **10-10 (above):** Once the curve has been laid out with the first pass, a saber saw will is used to remove the bulk of the wood. One more pass of the router finishes the arc.

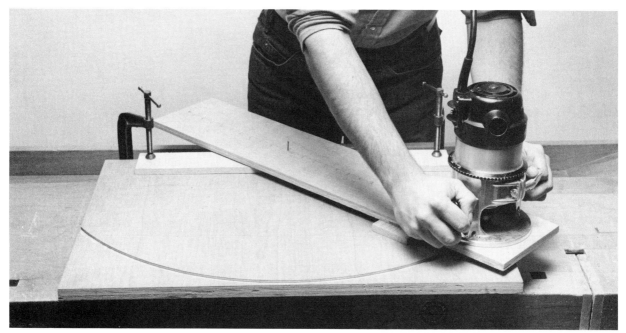

CURVED GROOVES

Decorative circular grooves, as shown in **figure 10-11,** can be cut using either a wedge-shaped guide on the router table, as shown in **figure 10-12,** or a circle-cutting jig.

10-11: Decorative curved grooves are made on the router table when you do not want a nail hole in the workpiece.

10-12: The decorative grooves in **10-11** were made on the router table with this wedged fence. When you keep the edge of the workpiece in contact with the inner faces of the wedged fence, the bit cuts in at a fixed dimension.

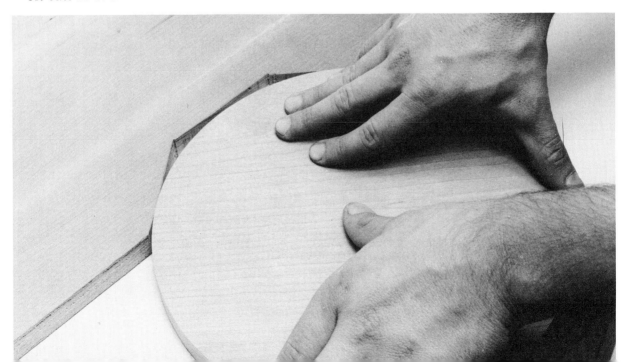

ROUTING ROUND STOCK

Routing round stock presents a particular challenge to the woodworker because it is unlike most other routing situations. Here the router cannot rest on the surface that is to be cut. In addition, supporting and clamping the workpiece so that it does not move while the cuts are being made can be a problem. To communicate the nature of these problems clearly, I'm going to discuss specific examples.

Fluting Straight and Tapered Cylinders

Fluting on a round cylinder is ideally done while the wood is still on the lathe. This will require careful layout, precise engineering, and a little experimentation to get the alignments just right. Turn an extra test piece for this purpose. Begin by cutting a strip of paper to the widest dimension of the stock (**fig. 10-13**). Divide this into the desired number of flutes. Ideally, your lathe will have an indexing head, which will be used to fix the position of your workpiece (**fig. 10-14**). You might want to adjust the spacing between the flutes to correspond with the indexing divisions. Remember, if the turning is ta-

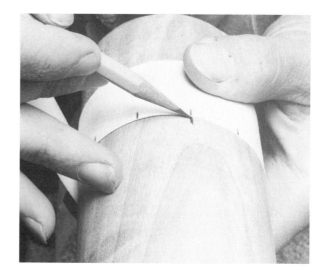

10-13 (above): Marking the flutes with evenly spaced marks on a strip of paper. **10-14 (below):** An indexing head divides the cylinder and locks the workpiece into position.

pered or curved, the flutes will be closer together at the narrowest point on the cylinder.

Now, to support the router, you will have to construct a lathe-routing fixture, sometimes called a fluting jig (**figs. 10-15** and **10-16**). The size has to be adjusted to your own lathe and workpiece sizes. Distance "d" between the two side pieces should be equal to the width of your router base. The router will rest on the A rails (**fig. 10-16**). Depth of cut will be controlled by raising and lowering the bit. Piece B will automatically center the fixture on the lathe bed. This step is critical. If the fixture is not perfectly centered, the flutes will not be straight. The jig is usually bolted to the lathe bed but on some lathes can be held with clamps. Stop blocks, marked C, are positioned to control the length of cut.

This fixture, as described, works perfectly well on straight cylinders but must be modified when the cylinder is tapered. In other words, the A rails must be on the same angle as the taper (**fig. 10-17**). The choice is either to angle the routing fixture or angle the rails. Angling the fixture is done with wooden wedges, but you can design the rails so that they can be pivoted to different angles.

Chuck the extra workpiece in the lathe and position the stop blocks. Rout the first flute. Move the indexing head and rout again. Continue. The space between the last and first flute should be exactly the same as all the other spaces. If it is not, you will be glad you experimented first on a spare piece.

Now, what about those of you who do not own a lathe? (I would like to know where your turnings came from!) You will have to construct a fluting fixture such as the one shown in **figures 10-18** and **10-19.** This fixture is also a good substitute for a lathe without an indexing head. The "live" (in this case it is not) and "dead" centers can be borrowed from an old lathe or purchased from a bolt turned to a 60-degree point. The indexing head you make yourself by cutting a round disc from ¾-inch (19-mm) wood and

10-15: The lathe-routing fixture.

drilling a series of evenly spaced holes. The holes should be of a size to accommodate a 6d nail, which is used to lock the indexing head and workpiece in a fixed position. The tailstock is adjustable, to accommodate workpieces of various lengths. The router support would be identical to the one used for fluting on the lathe.

10-16: Plans for a lathe-routing fixture. Exact sizes will vary, in relation to the size lathe and router being used. **10-17 (bottom):** The rails must be on the same angle as the taper.

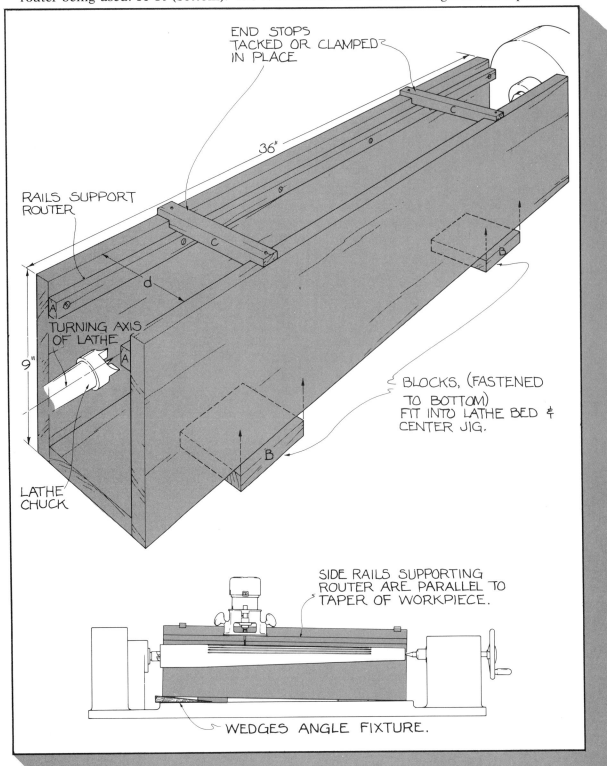

END STOPS TACKED OR CLAMPED IN PLACE

36"

RAILS SUPPORT ROUTER

d

C

A

TURNING AXIS OF LATHE

9"

A

LATHE CHUCK

B

BLOCKS, (FASTENED TO BOTTOM) FIT INTO LATHE BED & CENTER JIG.

B

SIDE RAILS SUPPORTING ROUTER ARE PARALLEL TO TAPER OF WORKPIECE.

WEDGES ANGLE FIXTURE.

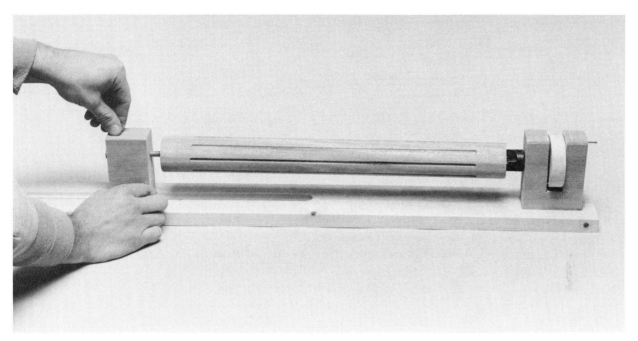

10-18 (above): If you do not have a lathe, this fluting fixture will work. **10-19 (below):** Plans for the above fluting fixture.

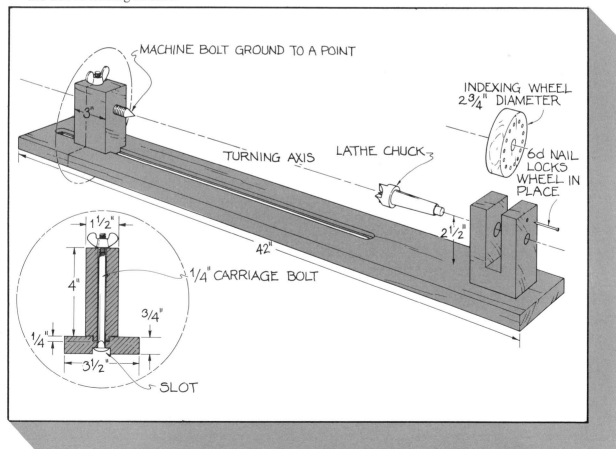

MACHINE BOLT GROUND TO A POINT

3"

TURNING AXIS

LATHE CHUCK

INDEXING WHEEL 2¾" DIAMETER

6d NAIL LOCKS WHEEL IN PLACE

42"

2½"

1½"

4"

¼" CARRIAGE BOLT

¾"

¼"

3½"

SLOT

Fluting Without a Lathe

Before I owned a lathe, I developed, out of necessity, another technique for fluting cylinders. Actually, it is an excellent system even if you do own a lathe. Any square block can be converted into an octagonal shape by making four rip cuts on the table saw. The fluting can now be done on the eight flats, as shown in **figure 10-20,** and the cylinder made round afterward. Do the fluting on the router table, using stop blocks. Before I owned a lathe, I used to shape the octagon into a cylinder by hand, but it is, of course, faster to do this on a lathe.

Fluting Curved, Tapered Cylinders

There are a number of different approaches to fluting curved, tapered cylinders, but the one I like best utilizes a ball-bearing pilot to

10-20: Routing flutes on the flats of an eight-sided piece is probably the easiest method for fluting. Stop blocks control the length of cut. The piece is turned round afterward.

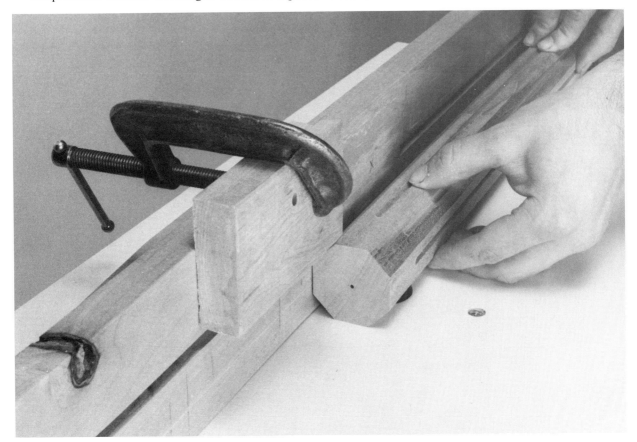

follow the contours of the workpiece (**fig. 10-21**). The cutting here is done laterally, rather than from above. To do this, you will need a specially made router bit. Most bit manufacturers (see *Sources of Supply*) can customize such a bit, but it will usually be expensive. The round-nose edge of the bit can be fashioned from a slotting cutter. A very large bearing will have to be used to control the shallow cut. You will also have to control the overarm jig (**fig. 10-21**). The bottom of this jig rests on a flat, smooth board attached to the lathe bed. At first, this type of routing will feel a little awkward, but so long as the workpiece is tightly secured and the bit is adjusted to the exact centerline of the workpiece, very little can go wrong.

The same bit, or any variation of it, can also be used to solve another routing problem that has always intrigued me. Consider the curved pedestal leg in **figure 10-22.** Fluting on a piece of this shape is usually done

10-21: Fluting a curved, tapered cylinder. This method utilizes a custom-designed router bit with a ball-bearing pilot. The router is guided with a specially made overarm sliding jig.

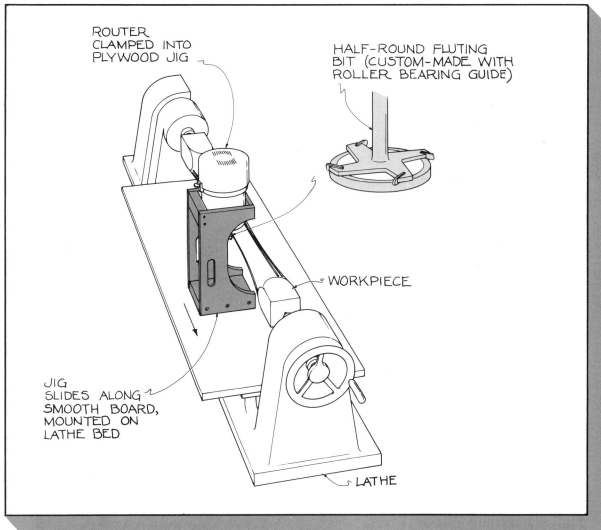

10-22: The same custom bit used in **fig. 10-12** can be used to solve another routing problem. The router rests on the flat of the workpiece and will follow its every contour.

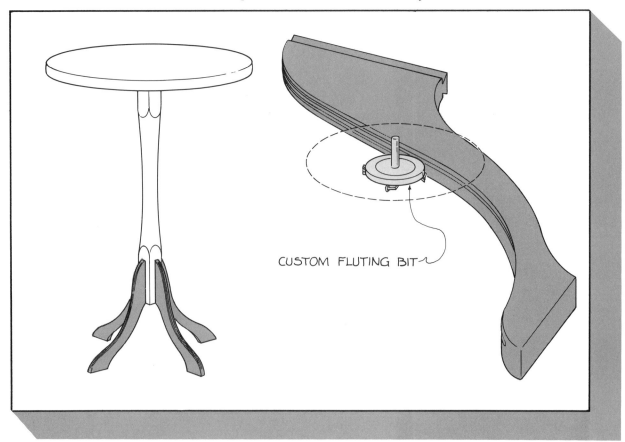

CUSTOM FLUTING BIT

on a shaper, but with a customized bit can easily be done with the router. The router rests on the flat and will follow any contour. Depth of cut can be controlled only by changing the size of the bearing.

The Slotted Dovetail

The technique for making a slotted dovetail provides a good example of how to cut a deep groove in round stock. Relatively deep grooving differs from fluting in that greater forces are being exerted on the wood and, in

return, on the router. As a result, it becomes more important that the wood be locked firmly in position. The workpiece can be affixed to the lathe, but added supports are necessary. An indexing head will also help to lock the workpiece in place. Another problem with this type of routing is that the cuts often have to be made near the end of the stock. The headstock can often get in the way of the router base; on narrow pieces, the router bit could actually come in contact with the live center. To prevent this from happening, cut the turnings oversize so that the parts to be routed are away from any possible interference (**fig. 10-23**).

An alternative to routing on the lathe is to build a routing fixture for each different turning you have to rout. This fixture is es-

10-23: Cutting the turnings oversized facilitates routing on the lathe. In this way, the router is well away from the headstock and tailstock.

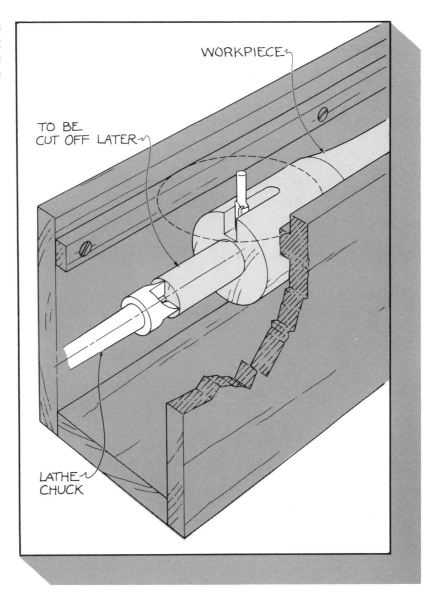

WORKPIECE

TO BE CUT OFF LATER

LATHE CHUCK

sentially a cradle that not only clamps and supports the wood but also provides a platform on which the router rests. There are many variations in the design of these fixtures. One way to hold the workpiece in position is to turn tenons at both of its ends. These tenons fit into round notches at either end of the routing fixture (**fig. 10-24**). When routing is complete, they are cut off. Another technique is to make the turning fit tightly into its cradle and tighten it further with screws (**fig. 10-25**). Additional support yokes or wedges are necessary in either technique.

To make a slotted dovetail for a tripod pedestal table, build a routing fixture like the one shown in **figure 10-26**. You could rout the dovetail directly into the turned post, but this would mean that the shoulders of each dovetailed leg would have to be shaped to conform to the radius of the turning. An alternative is for you to rout a flat on the post as your first step, as shown in **figure 10-26**. The flat should be only fractionally longer than the actual thickness of the leg.

10-24: One way of clamping a round workpiece.

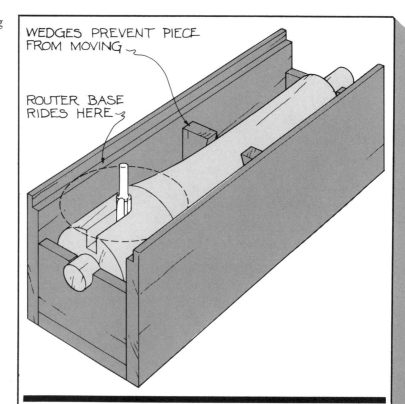

WEDGES PREVENT PIECE FROM MOVING

ROUTER BASE RIDES HERE

10-25: Another cradle for securing a round workpiece.

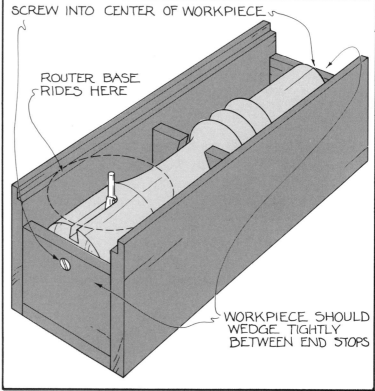

SCREW INTO CENTER OF WORKPIECE

ROUTER BASE RIDES HERE

WORKPIECE SHOULD WEDGE TIGHTLY BETWEEN END STOPS

 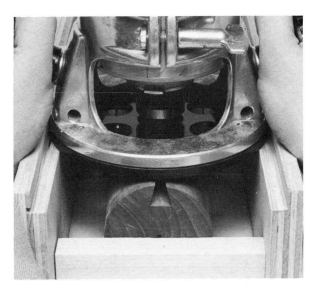

10-26 (left): The workpiece is held tightly in its length by the fixture itself, and in the width with wedges. An end stop limits the travel of the router. The cut you see is a flat, routed onto the round column where a pedestal leg will be dovetailed. **10-27 (right):** Cutting a slotted dovetail.

10-28: Slotted dovetails on a pedestal column.

10-29: Cutting the dovetail itself on the router table.

Dovetails have to be cut in one pass. The bit, by virtue of its shape, cannot be lowered in stages. This can put a strain on the router, particularly if the wood is a hardwood such as maple. Drilling out the bulk of the material before routing makes it easier for the router to perform. Finally, to control the length of the dovetail cut, use an end stop (**fig. 10-26**).

The same dovetail bit will be used to cut the dovetail on the leg. This is best done on the router table, since you would have a difficult time trying to balance the router on the leg itself (**fig. 10-29**).

11 Freehand Routing

Freehand routing is fun routing. No fences, no piloted bits, no template guides—just you and your hands to guide the router. Concentration is intense and a moment's lapse can ruin a cut. Freehand routing is used in a number of different ways, but mainly for lettering, inlay, surface texturing, and removing background wood for carving.

The degree of control you possess is largely a result of how deep you cut. When routing for inlay, where the depth of cut is usually less than $\frac{1}{32}$ inch (.79 mm), the router is surprisingly easy to control. Then, it is simply a question of the steadiness of your hands and your ability to concentrate. If you start cutting at a depth of, say, $\frac{1}{4}$ inch (6 mm), however, the router will have a tendency to wander. So, you are not only trying to rout along a line; you are also trying to control an unpredictable router. An explanation of why the router wanders is found in "The Physics of Routing" in Chapter 3, *Cutting Methods*.

As a general rule, it is best to rout in such a direction that if the router wanders, it will wander away from the line of cut rather than toward it (**fig. 11-1**). Make test cuts to see how the router responds to a given depth of cut in the particular type of wood you are working with. Some woods, such as redwood, are a woodworker's dream because of the ease with which they can be cut. Bass-

wood and pine are other fine choices for the carver and signmaker. But any wood can be routed freehand. Just rout more slowly when you encounter a swirling grain. Knots, in particular, can cause a router to jump.

In freehand routing, good lighting becomes even more important than usual. You will be viewing your pencil lines through the hole in the sub base, so a nearby source of light

11-1: Rout in such a direction that if the router is to wander, it will do so away from the line of cut rather than toward it.

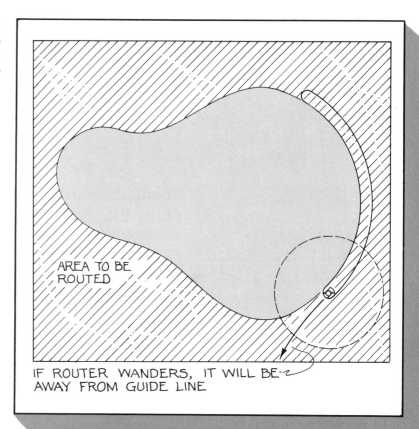

will be a big asset (**fig. 11-2**). I also suggest that you lightly pencil-in the areas to be routed. There is nothing so frustrating as routing a rose petal off its stem! In addition, make yourself an auxiliary sub base just for freehand routing, as shown in **figures 11-3 and 11-4.** This type of base will give you a much clearer view of the workpiece.

The position of your hands is also very important. Whenever possible, rest your elbows on the wood or workbench (**fig. 11-2**). Here, the movement of the router is controlled more with the hands and wrists than with the arms. Because you are peering in on an angle through the auxiliary sub base, you will find it easier to follow the pencil lines if you make your freehand cuts by bringing the router toward you.

11-2: In freehand routing, working under strong, direct lighting is very important. When possible, your elbows should rest on either the workpiece or workbench.

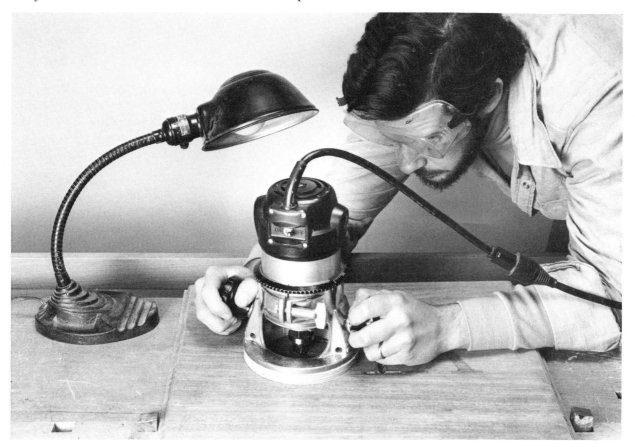

11-3 (left): An auxiliary sub base gives you a better view of the workpiece. **11-4 (right):** In this odd-looking photograph, a laminate trimmer with a flush-trimming bit is being used to trim an auxiliary sub base flush to the base. Since cutting edges of the bit come to within 1/16 inch (1.5 mm) of the metal base, perform this operation with great care.

What about the router itself? Certainly, the smaller the router, the easier it is to control. Considering this, I am surprised that more woodworkers do not take advantage of their laminate trimmers for shallow cutting and freehand routing. A laminate trimmer can be guided comfortably with one hand. In fact, with a trimmer, you can write your name in script almost as if you were using a pen. For deeper freehand routing, of course, you will need a moderate-weight router of at least 1 h.p. Rout away.

LETTERING AND SIGNMAKING

Before discussing freehand lettering, let us consider the alternatives. Several manufacturers make sets of letter and number templates that are used in conjunction with template guides. Sears, in particular, offers a wide variety of letter styles (Oriental, modern, Old English, script, and computer), in heights of up to 2½ inches (63.5 mm). For straightforward signs, these templates do a fast, clean job. Letter templates do have their limitations, however, since a specific sign will often require larger letters or a unique style of lettering. Template-made signs can also look somewhat stilted.

Another alternative is to purchase any of the several signmaking machines. These machines are intended for commercial use, though, and the better ones tend to be rather expensive. I would guess that most one-person sign shops operate without signmaking machines. Suffice it to say that most signmakers, if they are not carving by hand, probably rely on freehand routing. More likely, they employ both techniques.

To what degree the router can and should be used in lettering and signmaking is partly a matter of opinion, partly a question of the degree of refinement required. The fact is that a router, operated freehand, cannot cut letters with the same degree of precision as a fine carver can with a chisel. But on certain large signs (outdoor signs, for example), where the standards of perfection are relaxed and where less stylized letters are needed, the router is used to a greater degree, sometimes exclusively.

In routed signs, letters and designs are either incised (routed out of the wood) or the background is routed away so that the letters are raised. To rout large, incised block-style letters, begin by chucking a bit that is narrower than the narrowest part of the letter. If the narrowest part of the letter is 1 inch (25 mm), for example, use a ½-inch (12-mm) straight cutting bit to remove the bulk of the letter, then switch to a 3/16-inch (4.7-mm) or ¼-inch (6-mm) bit to rout along the pencil lines. In general, I set up one of my routers with the larger bit to do the rough work, then chuck the smaller bit in my laminate trimmer to finish the details. With the bulk of the letter routed with the large bit, the smaller bit need remove very little wood, making the trimmer very easy to control. The smaller bit can also get into much tighter corners than its heavyweight counterpart. You may also opt to use a round-nose bit to cut along the pencil line (**fig. 11-5**).

You will notice that with this type of routing, very small curls or shavings rather than sawdust surround the bit, often obscuring the pencil line. This buildup can be minimized by scoring the line with a razor blade or X-acto knife before you rout.

Rout the vertical parts of the letter first (**fig. 11-6**). If you want to, you can use a sim-

11-5: A round-nose bit can be used to soften the hard edges of a letter.

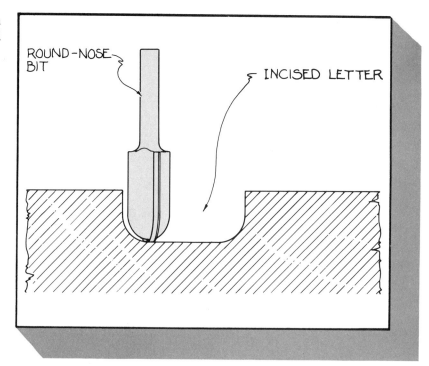

11-6: Rout the vertical parts of a letter first.

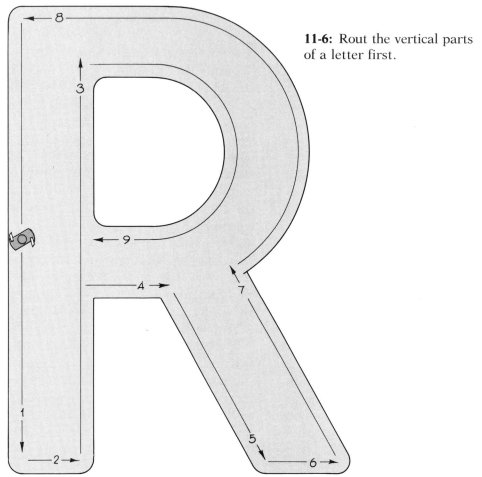

11-7: A T square (I use my dado jig) can be used to make any straight cuts.

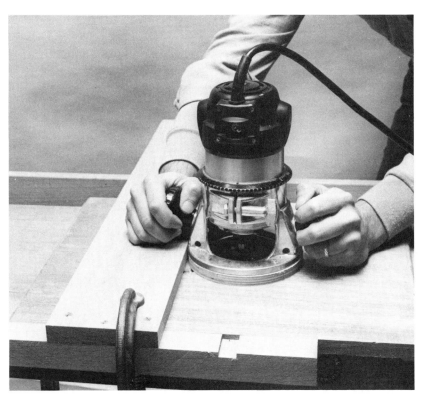

ple T-square arrangement to keep the lines straight (**fig. 11-7**). Work into the horizontal and curved sections of the letters. Always move the router from cut portions into uncut portions, to prevent chipping.

For Roman letters, I think that a misconception exists that all you have to do is chuck a V-groove bit and rout away. A Roman letter widens, tapers, and curves with infinite grace. Because it widens at various points, the angle of "V" would have to widen at these points too. What many signmakers choose to do, if the letters are very large, is to rout out portions of them and finish with carving tools (**fig. 11-8**).

Raised letters are formed by removing the background. You could make templates, but because most signs are different, it is probably easier (and faster) to rout the letters freehand. I begin by outlining the letters

11-8: When routing large letters, many signmakers use a router to rough out portions of the letters. Usually the letters are then finished with carving tools.

11-9 (left): Marc Bassett, a well-known woodcarver and signmaker in West Stockbridge, Massachusetts, uses the router extensively to rough out portions of his letters and designs, but always finishes the signs with hand tools. **11-10 (right):**   Although the textured surface behind the train could also have been obtained with a large core-box bit, Bassett did this with a carver's gouge.

with my carving tools (**fig. 11-11**). I then rout to within 1/16 inch (1.5 mm) of these cuts and finish with carving tools (**figs. 11-9, 11-10**).

If the piece has a symmetrical border, as shown in **figure 11-12,** use an edge guide to control the straightness of the cut. If it is a circular sign with a circular border, use the basic circle-cutting jig.

For removing large backgrounds, particularly where multiple depths are required,

I outfit my router with an oversized base (**figs. 11-13** and **11-14**). Make this base from a perfectly straight hardwood. It should be 7/8 inch (22 mm) thick, so as not to deflect from the length of its span and the weight of the router. This base is rested on a raised frame that surrounds the workpiece. This system facilitates maximum control, in that the router is independent of the workpiece, and multiple depths are easily regulated.

11-11 (left): Although it might not be necessary, I outline my letters with carving tools before beginning the freehand routing. **11-12 (right):** Instead of routing a symmetrical border freehand, it is easier to use an edge guide.

11-13 (above): To remove large backgrounds, use an oversized base. The base rests on a frame that is slightly larger than the workpiece. Wedges between the workpiece and frame prevent the workpiece from moving. **11-14 (below):** A better view of the auxiliary router base.

INLAY

The router is my choice for inlaying veneers, not only for its speed but also because it cuts to a precise and consistent depth. Let us first consider inserts. Marquetry inserts, sometimes called veneer assemblies, are available in a wide variety of shapes, patterns, and types of wood (**fig. 11-15**). Without being an expert, you can still bring the art of marquetry to your pieces. Inserts come with a paper covering that faces up. The paper is removed only after the insert has been inlaid.

Begin by aligning the insert on your workpiece and trace around it with a sharp pencil. Remove the insert and make an incision along the inner edge of this line with an X-acto knife or razor blade. Or, use two small pieces of masking tape to hold the insert down and to trace the outline directly with the X-acto knife (**fig. 11-16**). Marquetry inserts are typically ¹/₂₈ inch (.76 mm) thick. The depth of cut should be fractionally less than this. Make test cuts. Rout as close to this line as feels comfortable. Clean up the recess with either the X-acto knife or your carving tools.

You can also make your own inserts. First, make a template of the shape you want from ¼-inch-thick (6-mm) wood. Trace the outline of this design onto the veneer and roughly bandsaw it to shape. Using two-sided masking tape, attach the veneer to the template (**fig. 11-17**). After positioning your router on the router table, chuck a flush-trim bit. Trim the veneer to the final shape (**fig. 11-18**). Repeating this shape is now very easy.

Prefabricated inlay border strips are also

11-15 (left): A marquetry insert being separated from its background. **11-16 (right):** Inserts have a tendency to curl slightly. I press them into position with masking tape and trace directly around them with an X-acto knife. (I purchased these inserts from Constantine's.)

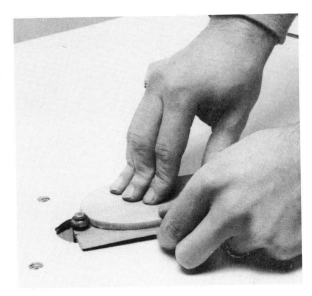

11-17 (left): When making your own inserts, attach the veneer to the pattern with masking tape.
11-18 (right): Trimming the veneer with a flush-trim bit. Notice that I am feeding the wood in the "wrong" direction, according to the conventions of routing. Veneer has a tendency to splinter, but routing and feeding in the direction shown will prevent this.

11-19: Inlay border strips are being set into a routed groove.

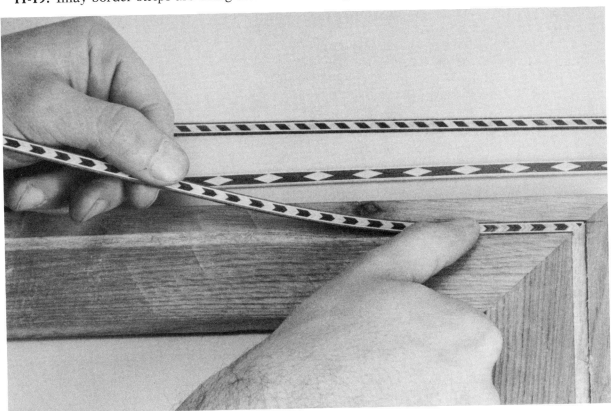

11-20: This box, built by my coworker Jeff All, has a top that was textured with a round-nose bit.

commercially available. These strips, which can be quite dazzling, come in standard widths from ¹⁄₁₆ inch to 1 ½ inches (1.5 to 38 mm). Use a standard edge guide and the appropriate size bit to make the channel, then glue the strips in place (**fig. 11-19**).

The backgrounds of carvings and signs can be enhanced by texturing the surface. Different pilotless bits will produce radically different effects. A small round-nose bit, for example, will give the impression that the background was finished with a small carver's gouge (**fig. 11-20**). A large core-box bit will simulate a larger gouge. A V-groove bit, run in perpendicular directions, will give the appearance of a stamped background (**fig. 11-21**). Experiment with your pilotless bits.

11-21: The stamped effect on this cherry board was produced with V-groove bit and router table.

12 Overhead & Pin Routing

Just as the router can be mounted from below, as with a router table, it can also be mounted from above. A router mounted from above is known as an overhead (or overarm) router.

One of the chief advantages of an overhead router is that the cutting action is always in full view. This is particularly helpful in freehand routing. In addition, overhead routers feature a foot-controlled lever that elevates the worktable and workpiece into the spinning bit. This permits you to make flawless beginnings and ends of cuts. Usually, when speaking of an overhead router, one refers to a fully integrated, self-contained industrial machine (**fig. 12-1**). Directly beneath the centerline of an overhead router's collet is a pin, which is mounted on the table below. It is the use of this pin as a guide that enables the machine to perform so many different operations. This use of the pin also explains why overhead and pin routing are virtually synonymous terms.

Pin routers are used throughout the woodworking industry primarily as production and duplication machines. Except for larger commercial shops, these machines are way beyond the needs and pocketbooks of most woodworkers. Yet there are ways of converting a portable router into an overhead or pin router, thereby increasing its versatility even further.

12-1: An industrial pin router. (*Ekstram, Carlson & Co.*)

MOUNTING THE ROUTER

The first question in adapting a conventional router to overhead use is how to physically mount the router. I have seen many homemade and machine-made clamping devices, some done more successfully than others. Not only is it imperative that the router be locked firmly in position without the possibility of movement, but the router must also present a bit that is exactly perpendicular to the work surface.

One effective method is to attach your router to your radial arm saw (**fig. 12-2**). Although you can make your own yoke for this purpose, I would suggest buying a ready-to-mount bracket. One company, HIT Distributors (see *Sources of Supply*), makes an excellent (and relatively inexpensive) mounting bracket that will fit most routers to most radial arm saws. The greatest problem with homemade overhead routers is how to lower the bit safely into the workpiece. You certainly cannot let go of the workpiece while you are fiddling with knobs and clamps. A bracket system mounted on the radial arm

12-2 (left): A router mounted to a radial arm saw. In this photo, the wood is moved with the router stationary. **12-3 (right):** Making an angled, overhead cut with a router mounted on a radial arm saw. (*Photos courtesy of HIT Distributors, Inc.*)

12-4 (left): The Shopsmith Router Arm. **12-5 (right):** The Shopsmith Router Arm can be mounted directly to a drill-press column.

saw takes advantage of the saw's raising and lowering mechanism. One hand holds the workpiece while the other controls the depth of cut. Once the bit is cutting at the proper depth, it can be locked in place easily, and both hands are free to guide the workpiece.

Mounting the router on the radial arm saw also enables you to take advantage of the saw's pivoting arm, so that you can make arcs of various radii. Another feature of a saw-mounted router is in making repetitive grooves and dadoes. The router can be manipulated in the same way that the saw would naturally be used. This is particularly handy when the grooves and dadoes have to be stopped. Finally, the yoke of the radial saw can be changed from its vertical position to produce angle cuts, which would be virtually impossible to do with other router setups (**fig. 12-3**).

You may also choose to purchase a self-contained pin router unit, such as the Shopsmith Router Arm (**fig. 12-4**). (Sears makes a pin-router unit as well.) The Shop-

smith Router Arm will accept any router motor with a diameter of between 3 inches and 4½ inches (7.62 cm and 11.43 cm). It features a quill feed handle that, with a twist of the wrist, allows the router to be raised or lowered to a predetermined depth of cut. The unit comes with an adjustable table and guide pins of ¼ inch, ⅜ inch and ½ inch (6 mm, 9.5 mm and 12 mm). This is an extremely well-made piece of equipment, designed with high safety standards. If you own a drill press and want to save a little money, you can purchase the arm separately without the column and table (**fig. 12-5**).

The inverted pin router, a combination of the router table and the concept of the pin router, is yet another possible approach. Your router is attached to the router table in the usual way and a hardwood or metal arm is extended over the table (**fig. 12-6**). A metal pin (bolt or shaft) is fitted through the arm with its centerline directly above the centerline of the router's collet. Quick alignment can be obtained by lowering the pin

12-6: A homemade inverted pin router. The pattern is guided along the template above while the cutter duplicates the pattern in the workpiece below.

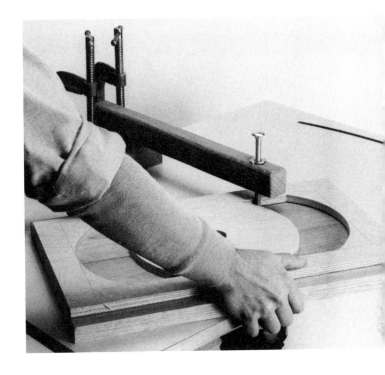

into the collet itself. This type of arrangement has certain advantages over a homemade overhead router. For one thing, the router is locked in position without the possibility of movement. Second, you can watch the template as you guide it along the pin, although some might argue that they would prefer to see the cutting action. Finally, there is a safety factor—the bit is not exposed. It is buried in the workpiece.

The arm must be rigid but easy to remove. The pin can be a ¼-inch (6-mm) bolt threaded through a T nut (**fig. 12-7**). Since the table cannot be elevated, you will have to lower your workpiece onto the protruding bit. The pin is then lowered through the arm and into or alongside the template.

Whichever system you use, be ever-vigilant in terms of safety. Pin routing, especially when the pin router is of the homemade variety, is potentially more dangerous than conventional routing techniques. Your fingers should *never* be held too near the bit. Wood fed in the wrong direction could be

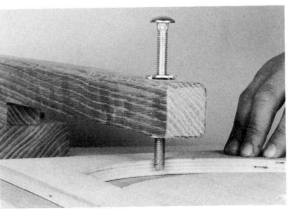

12-7: The arm of my inverted pin router is made from solid oak. The "pin" in this case is a ⅜-inch (9.5-mm) bolt threaded through two T nuts. (The lower T nut is not visible.) Using two T nuts gives the pin greater rigidity,

grabbed by the bit, thus pulling your hands toward the cutter. One feature of the Shopsmith Router Arm that I especially like is the full-view safety shield. Even if the bit were to grab the wood, your fingers would be kept away from the bit.

USING AN OVERHEAD ROUTER

Straightline Fence Routing

Straightline fence routing, in which the wood is advanced along a fence and into the cutter, is used primarily to make grooves, rabbets, mortises as shown in **figure 12-8**, and molded edges.

In terms of feed direction, the standard axiom in overhead routing is to feed the work against the rotation of the bit. This advice is somewhat misleading. **Figure 12-9** shows how you could follow this rule and still be feeding improperly: the wood is fed against the rotation of the bit, but the workpiece

12-8: Routing a mortise on an overhead router. A featherboard helps to keep the workpiece against the fence. Note the direction of feed.

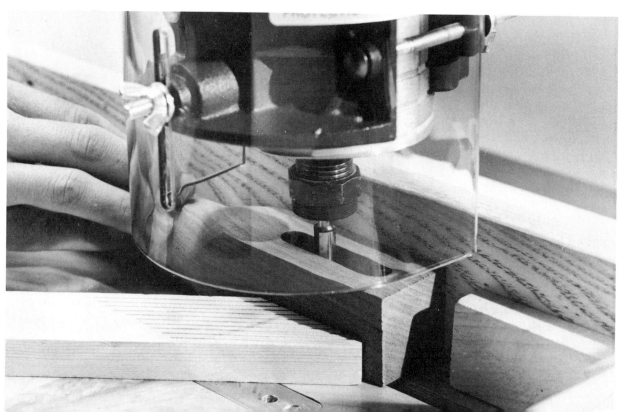

would be pulled away from the fence. **Figure 12-10** shows the proper direction of feed, which is the equivalent of left-to-right routing with a moving router. (See "The Physics of Routing" in Chapter 3.) A corollary, then, is that when routing the edge of a board, the routed edge should be against the fence and not between the fence and bit.

Straightline fence routing opens up a broad range of possibilities in the making of moldings (**fig. 12-11**). Any number of different pilotless bits can be used in combination

12-9: Incorrect feeding. Although the wood is being fed "against" the rotation of the bit, as it should, routing with the wood between the fence and bit will result in the wood being pulled away from the fence. **Safety note:** To allow this photo and those in **12-10** and **12-11**, the recommended safety shield was removed.

12-10: This is the correct direction of feed in straight-line fence routing.

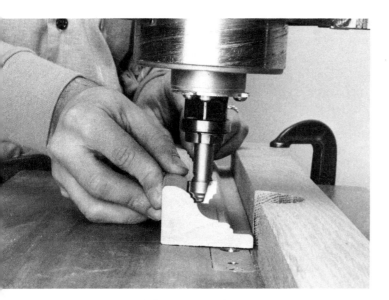 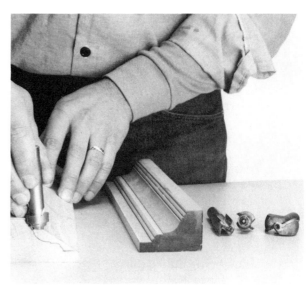

12-11 (left): An overhead router can cut large and intricate moldings. **12-12 (right):** The moldings you make are often dictated by the bits you have available. First, lay out the profile you want and find the bits that most closely correspond. Note, too, that much of the stock at left was first removed by a table saw. At right, the finished molding.

to make profiles that would be impossible to form on the router table. Whenever it is possible, cut the moldings on much wider stock than the final desired shape. This will give you safer control of the cut, since the workpiece will be better balanced and more rigid. Any large amounts of wood to be removed (wasted) should, if possible, be cut on the table saw (**fig. 12-12**).

Contour-Edge Shaping

In contour-edge shaping, the edges of curved or irregularly shaped workpieces are fed against a piloted bit on a guide pin (**fig. 12-13**). If the complete edge of a piece is to be shaped, as shown in **figure 12-14**, the guide pin would ride against a template.

Guide pins can also be used to shape curved moldings. The use of different bits is the same as with straightline fence routing, except that the workpiece is riding between two collars (**fig. 12-15**). Cut your moldings several inches oversize in length, since irregularities often occur at the beginnings and ends of cuts. The collars (shaper collars work fine) fit over dowels that have been glued to an auxiliary worktable. The distance between collars should be fractionally larger than the width of the molding. Be particularly careful at the beginning of the cut. The workpiece must first be held against the back collar and pivoted into the bit. Failure to do so could result in kickback. The wood is fed along the back collar, but it is the front collar that keeps the wood in a consistent position. Without the front collar, the back collar would become a fulcrum, in a sense—any lateral movement in the feed would ruin the cut. In other words, the front collar prevents the workpiece from pivoting.

12-13: The contoured edges of the workpiece ride against a pin while the bit cuts the desired shape.

12-14: If the complete edge is to be shaped, a template should be added to ride along the guide.

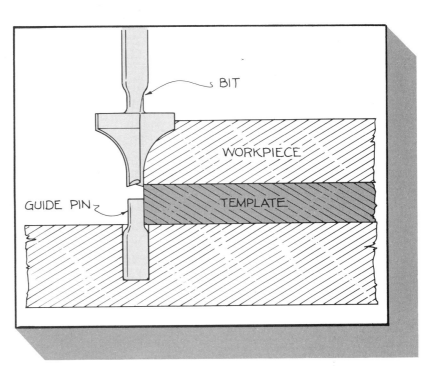

12-15: When routing curved moldings, feed the stock feed between two collars. Note the direction of feed.

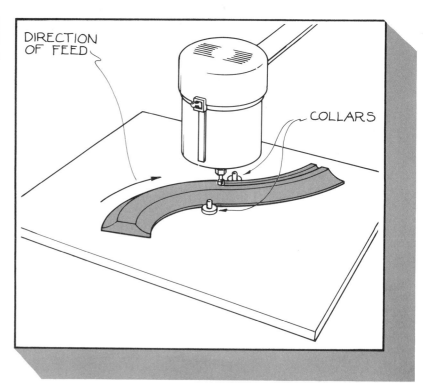

Duplicating with Patterns

Duplicating, through the use of templates and guide pins, is the process most commonly associated with overhead routing. Let us first consider decorative grooving (**fig. 12-16**). The concept is simple. To make the template, begin by laying out the design on a piece of plywood, masonite, or other material. Cut, shape, and sand this piece. Screw it to another blank (**fig. 12-17**). Using a pin and a bit of the exact same size, place the design face down and guide the pattern against the pin. The bit will reproduce the

12-16: A template below rides over the guide pin while a cutter duplicates the pattern above. (*Photo courtesy of Shopsmith, Inc.*)

12-17: Screwing the blank to another piece, which will ultimately become the working template.

12-18: Internal excavation. The pattern rides along the template while the bit reproduces the desired shape. Changing to a different diameter pin will change the size of the opening. This is sometimes done to produce a rabbeted recess.

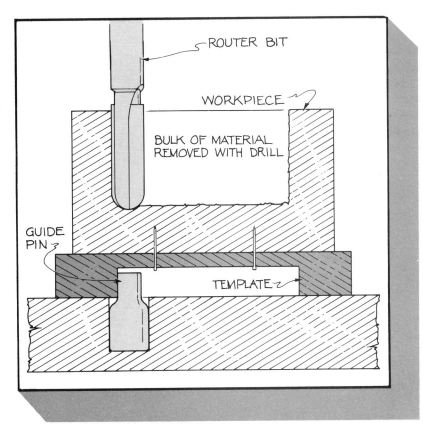

12-19: How to shape the edges of the recess.

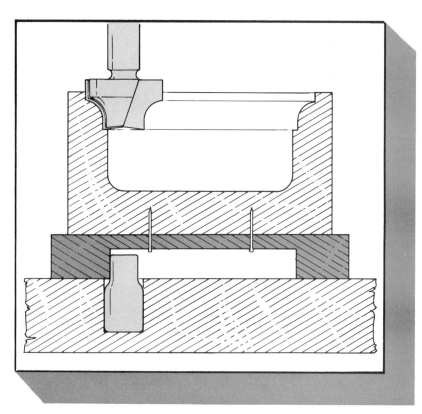

12-20: A typical duplicating setup. (*Photo courtesy of Shopsmith, Inc.*)

desired shape on the top piece. This piece now becomes the template for future duplication. The template with the desired shape grooved into it is fit over a guide pin. The bit is lowered into the workpiece. The pin guides the template, and the shape of the template is transferred to the workpiece.

The same concept is used for internal excavation and shaping. **Figure 12-18** shows a typical setup. When excavating large and deep areas, drill out as much stock as possible, or make the router cuts in several passes. Shaped pilotless bits are also used to shape the edges of the recesses (**fig. 12-19**).

SOURCES OF SUPPLY & METRIC CONVERSIONS

Router Manufacturers
Black & Decker
Bosch (formerly Stanley)
Hitachi
Makita
Milwaukee
Porter-Cable (formerly Rockwell)
Ryobi
Sears

Industrial Overhead/Pin Routers
Ekstrom, Carlson & Co.
P.O. Box 1611
Rockford, IL 61110

Onsrud Overhead Routers
Danly Machine Corp.
2100 South Laramie Avenue
Chicago, IL 60650

Overhead Router Arms and Attachments
HIT Distributors
2867 Long Beach Road
Box 535
Oceanside, NY 11572

Shopsmith Inc.
750 Center Drive
Vandalia, OH 45377

Router Bits
Dinosaw Inc.
340 Power Avenue
Hudson, NY 12534

Ekstrom, Carlson & Co.
1400 Railroad Avenue
P.O. Box 1611
Rockford, IL 61110

Forest City Tool Co.
P.O. Box 788
620 23rd Street, N.W.
Hickory, NC 28603

Fred M. Velepec Co., Inc.
71–72 70th Street
Glendale, NY 11385

Greenlee Tool Division
2330 23rd Avenue
Rockford, IL 61101

Porter-Cable Corp.
P.O. Box 2468
Jackson, TN 38301

Dovetailing Accessories

David Keller
Star Route
Box 800
Terrace Avenue
Bolines, CA 94924

Leigh Industries Ltd.
Box 4646
Quesnel, British Columbia
Canada V2J 3J8

Marquetry Accessories and Lumber Supplies

Albert Constantine & Son Inc.
2050 Eastchester Road
Bronx, NY 10461

Sharpening Services

Dinosaw, Inc.
340 Power Avenue
Hudson, NY 12534

Metric Conversion Table

Inches	Millimeters
1/16	1.58
1/8	3.17
3/16	4.76
1/4	6.35
5/16	7.93
3/8	9.52
7/16	11.11
1/2	12.70
9/16	14.28
5/8	15.87
11/16	17.56
3/4	19.05
13/16	20.63
7/8	22.22
15/16	23.81
1	25.40

Index